T0305765

THEORETICAL FOUNDATIONS FOR QUANTITATIVE FINANCE

THEORETICAL FOUNDATIONS FOR QUANTITATIVE FINANCE

Luca Spadafora
Università Cattolica del Sacro Cuore, Italy

Gennady P Berman
Los Alamos National Laboratory, USA & New Mexico Consortium, USA

World Scientific

NEW JERSEY · LONDON · SINGAPORE · BEIJING · SHANGHAI · HONG KONG · TAIPEI · CHENNAI · TOKYO

Published by

World Scientific Publishing Co. Pte. Ltd.

5 Toh Tuck Link, Singapore 596224

USA office: 27 Warren Street, Suite 401-402, Hackensack, NJ 07601

UK office: 57 Shelton Street, Covent Garden, London WC2H 9HE

Library of Congress Cataloging-in-Publication Data
Names: Spadafora, Luca, author. | Berman, Gennady P., 1946–   author.
Title: Theoretical foundations for quantitative finance / by Luca Spadafora,
    Università Cattolica del Sacro Cuore, Italy, Gennady P. Berman,
    Los Alamos National Laboratory, USA & New Mexico Consortium, USA.
Description: New Jersey : World Scientific, [2017] |
    Includes bibliographical references and index.
Identifiers: LCCN 2016055368 | ISBN 9789813202474 (hc : alk. paper)
Subjects: LCSH: Finance--Mathematical models. | Investments--Mathematical models.
Classification: LCC HG106 .S628 2017 | DDC 332.01/5195--dc23
LC record available at https://lccn.loc.gov/2016055368

British Library Cataloguing-in-Publication Data
A catalogue record for this book is available from the British Library.

Desk Editors: Dipasri Sardar/Alisha Nguyen

Typeset by Stallion Press
Email: enquiries@stallionpress.com

Printed in Singapore

Preface

The last decade was characterized by several financial crises that dramatically changed the whole financial framework, the behavior of market participants and, in general, the perception of financial investments. As a consequence of this sort of revolution, different cornerstones of the classical Quantitative Finance (QF) framework have been changed by more modern approaches that can better describe such a different financial system.

This book presents both the classical and modern QF topics in a unified framework. It discusses which of the old paradigms are still valid in this new financial world and which have been reformulated. This coherent introduction to modern QF theory is addressed to readers with a strong quantitative background who seek to become employed in this field as Quantitative Analysts (or Quants) and to active practitioners interested in reviewing the recent changes in QF.

The book deals with quite a wide class of models, ranging from the rational pricing of financial derivatives to quantitative risk estimation. For this reason, it is a broad *introduction* in the sense that it is not focused on a particular topic, but aims to summarize the technical tools required in a wide range of QF topics. The applications described in the book, although close to professional models, should be regarded as summaries of concepts that show how a mathematical tool can be exploited in practical situations. For this reason, this book is well suited for junior professionals who want to learn more advanced topics that can be utilized in a variety of practical situations during their future careers.

On the other side, this book does not deal with issues related to optimal investment strategies and the forecasting of financial trends, even if we think that most of the mathematical tools and models described in the book can be considered as a good starting point for these kinds of applications.

Most of the topics in this book are derived from the lectures of a QF course given by one of the authors to graduate students at Catholic University of the Sacred Heart, Department of Mathematics, Physics and Natural Sciences, Italy; examples come from practical experience in the main Italian banks.

The authors gratefully acknowledge the assistance of E. Benzi, G. D. Doolen, M. Dubrovich, G. Giusteri, A. Marziali, A. Pallavicini, N. Picchiotti, A. Spuntarelli and M. Terraneo, who carefully reviewed the manuscript and provided many corrections and suggestions for its improvement. The authors thank N. Balduzzi for providing the illustrations.

Any remaining flaws are the responsibility of the authors.

L. Spadafora and G. P. Berman

Contents

Chapter 1

Introduction

Quantitative Finance (QF) is a discipline that lies between hard sciences, from which a quantitative approach is taken, and finance that refers to the field of economic modeling. In other words, QF is a field that aims to develop mathematical tools that contribute to a rational description of financial markets, for example, for the estimation of the risk of an investment or of the estimated price to be attributed to a financial instrument. This rational approach is somehow in contrast with the usual perception of financial-related activities that may look completely chaotic and non-scientific. From this perspective, a rational approach to this field could be understood, more than a utopia, as a regular subject with practical implications because of the multitude of hypotheses to be introduced. On the other side, we think that this rational approach is somehow needed as a way to facilitate the communication between financial stakeholders, starting from the commonly accepted rational cornerstone.

In the last decades of the 20th century and in the initial years of the 21st century, many new models for different objectives were developed, with increasing complexity from the mathematical point of view, that contributed to the coherent framework that we call *Classical Quantitative Finance* framework. Unfortunately, the situation dramatically changed in August 2007, when Lehman Brothers's firm closed its subprime lender (BNC Mortgage), several crises (e.g. Sovereign crisis in 2011) came in succession, and the financial world understood that its paradigms had to be modified. Nowadays, after about 9 years since that financial crisis, some classical paradigms of QF were reformulated, and a new *Modern Quantitative Finance* framework has been established.

This book introduces the classical and modern QF topics in a unified framework, discussing which of the old paradigms are still valid in this new financial world and which ones need to be reformulated. The book is organized as follows.

In Chapter 2, we introduce the main mathematical tools used in the following chapters to develop financial models. In particular, it focuses on Probability Theory and Stochastic Calculus. As these topics are quite advanced, a formal description for them would require extensive mathematical review; on the other side, a very high-level introduction (as done in many books on QF) would not give the readers the opportunity to *control* the mathematical tools and to develop models on their own. For this reason, we preferred to present a formal treatment of the topic, with only an intuitive description of the mathematical tools, explaining why each tool is necessary and what can be done with it. In this respect, we observe that the typical questions asked for Quantitative Analyst positions are usually based on these theoretical tools; for example, interview questions are often raised concerning Itô's integral definition.

In Chapter 3, we introduce the pricing problem and the main objectives of our models. This aspect is of fundamental importance in our opinion as we have observed that, in many cases, students and also practitioners have a wrong perception of what a pricing model should be able to do and what is not required. In addition, we present a very simple and robust method to price financial instruments, starting from a few hypotheses about the financial markets. In spite of its simplicity, this method is used in practice in many situations and, when applicable, it is usually preferred to more sophisticated models. Finally, in this chapter, we introduce the main financial instruments we consider in the rest of the book.

Chapter 4 contains the description of a very general algorithm to price a financial derivative that represents an alternative approach to the one presented in Chapter 3. Applications of these algorithms are presented in Chapter 5 where Black–Scholes (BS) model and its generalization are discussed. In this chapter, two demonstrations of the same model are presented: one based on a formal approach and one on a more intuitive description. The resulting BS formula is then exploited to obtain a more general framework that is closely related to the models used in practice by professionals. By simple pricing problems, different modeling techniques, such as the change-of-measure and the use of stochastic volatility are presented.

In Chapter 6, we consider the other fundamental topic of QF, i.e. the risk modeling. In particular, we focus on the market risk, and we show different methodologies to estimate the risk of an investment in a rational way. In the same chapter, we also discuss different statistical effects that characterize extreme events and how it is possible to test the performance of a risk model in practice.

Finally, in Chapter 7, we discuss how the classical QF framework can be extended to take into consideration the new context of the modern QF, giving different examples on how pricing formulas have to be changed and risk models have to be adjusted.

The details of derivations of some expressions are presented in Appendix.

Chapter 2

All the Financial Math You Need to Survive with Interesting Applications

2.1. Introduction

In this chapter, we are going to discuss the main mathematical tools that are used in the Quantitative Finance (QF). As the reader could figure out, this field can be potentially huge, and entire books have been written on this [1–3], and the interested reader can find therein all the mathematical details to develop a deep knowledge about the topic.

In this chapter, we would like to mention only the key aspects that can help the reader survive throughout the book, avoiding technical demonstrations or excessive formalism, and following the lines of Ref. [1]. The price to pay for this approach is a knowledge that is mainly based on intuition rather than on a formal basis that could be developed after a general understanding of the topic. On the other hand, we think that the intuition will help the readers develop an understanding of the general framework, allowing them to easily make connections between topics that could seem unrelated at first sight.

2.2. Probability Space

The first mathematical concept we need to introduce is the idea of probability and probability space. This aspect is of fundamental importance for the models we are going to introduce in the following chapters as we are implicitly assuming that the system we want to model, i.e. the financial market, can be efficiently described by the events subject to the probability law.

Even if this choice could seem quite natural for practitioners, it should be mentioned from the very beginning that this is an arbitrary choice that will condition the development of the whole mathematical framework for the system representation. In particular, we are going to assume that our knowledge about the system is somehow influenced by uncertain events that cannot be known exactly; as a result, the models we are going to deal with implicitly incorporate this uncertainty in their mathematical formulation.

In general, people have quite an intuitive understanding that the probability concept is strongly related to the idea of frequency. From a classical point of view, probability was defined as the ratio of favorable outcomes to the total number of possible outcomes, assuming that the latter have all the same probability. This definition is quite simple, but unfortunately has some drawbacks:

- It is a circular definition, i.e. we used the concept of the "same probability" to define the probability itself.
- It does not tell anything when events do not have the same probability.
- It works only if the total number of possible outcomes is finite.

In order to avoid these issues, some generalization is needed, requiring the formulation of a theoretical framework based on the measure theory. Actually, it is quite natural to define the probability as a *measure* of how likely an event is to occur. From this perspective, we observe that, in order to obtain a formal probability framework, we need to specify three fundamental elements:

- what we consider to be our whole *set* of events,
- all the combinations of events we consider of interest and we want to measure,
- *how* we want *to measure* the probability of an event.

The first element is quite natural. If we want to talk about the probability that an event occurs, we should be able, at least, to specify the events among which we want to move, the ones we want to describe and measure. We define this set as Ω, and we will call it a *set of scenarios*, meaning that events are properly defined subsets of Ω.

It is quite natural to expect that if one is able to measure two elements separately, one should also be able to measure the union of these two elements. In a probabilistic sense, if one is able to measure how likely two events are separately, it is quite natural to expect that one should be able to say how likely they are together. This requirement could hide some difficulties from a mathematical perspective if the set of events we are going to consider is infinite or uncountable. In this case, the union operation could lead us to some pathological case, where the union of two events (that are Ω-subsets) does not belong to the set of scenarios that we can measure. In order to avoid this mathematical issue, we need to define a proper collection of sets that would assure that these pathological cases are avoided. This is exactly the role of the *sigma-algebra* $\mathcal{F}$ (also denoted as σ-algebra) that is defined as a collection of Ω-subsets such as

$$
\begin{aligned}
&\emptyset \ \in \mathcal{F} \\
&\text{if} \quad A \in \mathcal{F} \Rightarrow A^c \in \mathcal{F} \\
&\text{if} \quad A_n \in \mathcal{F} \Rightarrow \bigcup_{n \geq 1} A_n \in \mathcal{F} \quad \forall \, (A_n)_{n \geq 1} \text{ disjoint.}
\end{aligned} \tag{2.1}
$$

In other words, the requirement of closure under the union operation poses some constraint on the possible collections of subsets of Ω that can represent our set of events; for this reason, we require that the set of events is the σ-algebra over Ω. By requirements (2.1), it follows that the empty set, $\emptyset$, and the set of the scenarios $\Omega = \emptyset^c$ ("not in $\emptyset$") always belong to the σ-algebra, where A^c is the complement of the set A.

The last element for our probability framework is a way to measure how likely is an event in our σ-algebra. In order to do this, we need to define a *probability measure* as a function $\mathbb{P} : \Omega \to [0, 1]$ such that $\mathbb{P}(\Omega) = 1$, and if $A_1, A_2, \ldots \subseteq \mathcal{F}$ is an infinite sequence (of disjoint subsets),

$$
\mathbb{P}\left(\bigcup_{n=1}^{\infty} A_n \right) = \sum_{n=1}^{\infty} \mathbb{P}(A_n). \tag{2.2}
$$

If all the requirements are satisfied, we call $(\Omega, \mathcal{F}, \mathbb{P})$ *probability space*. By these three simple elements, we can start to build our fully coherent theoretical framework. Here, we stress that, from this theoretical framework, the probability is not defined anymore as a ratio between positive outcomes of many experiments and the total number of the experiments themselves, as in the classical probability approach, but it is something more general that aims to measure how much likely an event is. In this respect, we observe that the probability associated to an event is not an absolute concept, but actually depends on the function $\mathbb{P}$ we choose to perform the measure and on the composition of the set of the measurable elements, i.e. the σ-algebra.

From a purely mathematical perspective, this framework belongs to the more general *measure theory* context, and many aspects of our simple framework could be analyzed at a very fundamental level. Given the goals of this book, we are not going to discuss these topics, and we will try to keep things as simple as possible.

What we want to mention here is that under suitable conditions, the probability measure can be represented as an integral over the events of interest, i.e. the following relation holds:

$$\mathbb{P}(\omega \in A) = \int_A d\mathbb{P}(\omega), \qquad (2.3)$$

where A is a generic subset of the σ-algebra. By this representation, the probability of an event can be estimated by standard integral calculation, and the measure related to each element ω is equal to 0. Because of this property, this kind of measure is called *diffuse*, and it is the main measure representation that we are going to consider in the rest of the book.

As a general remark, we observe that, from a mathematical perspective, the role of the measure $\mathbb{P}$ in the integral of Eq. (2.3) is to make the element ω "smooth enough" to be integrated by standard integral definition. In this respect, by the use of the probability measure $\mathbb{P}$, we can be sure that, no matter what the behavior of ω is, we can straightforwardly perform the integration. In Section 2.13, we will discuss how to perform the integration when it has to be performed with respect to some function of the element ω. In addition, we observe that ω should be understood as a random event that, in general, we do not know with absolute certainty. For this reason, ω is typically interpreted as a *noise* that we include in our model because of the lack of complete information about the system.

2.2.1. *Probability Measure and Random Variables*

In this section, we want to make more evident the relation between our theoretical framework, i.e. the probability space $(\Omega, \mathcal{F}, \mathbb{P})$, and the typical real life probability problems, in particular the ones related to QF world. First, we observe that in typical real life problems, we have to deal with observations that we believe behave according to a probabilistic law. For example, in QF, we have historical series of prices that could be understood as the numerical representation of our observation on the system we are interested in, i.e. the financial market. On the other hand, our theoretical framework generally deals with sets and measures that could seem to be too abstract. Of course, we could reduce the generality of our theoretical framework and simply assume that Ω is represented by the real numbers set $\mathbb{R}$, so that our generic event is actually the number related to the observation. This assumption strongly simplifies our theoretical framework, and it is usually considered as the starting point for an elementary introduction to probability theory.

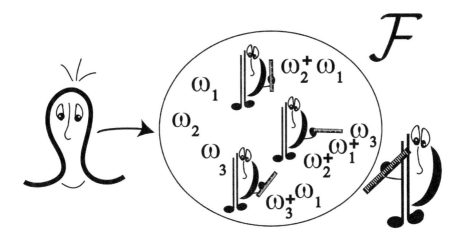

Fig. 2.1.　A representation of the Ω space and the role of the probability measure.

In order to keep things more generic, we could introduce a function X that *maps* the generic event $\omega \in \Omega$ into the set of real numbers $\mathbb{R}$; in this way, we split the concept of random event ω that represents the source of noise in our system and the result of our observation $X(\omega)$. In this approach, we can think of the financial market as a place where different sources of randomness (e.g. news, investor opinions, political decisions, etc.), represented by the state variable ω, generate at each time new prices $X(\omega)$. By this simple approach, we are able to discern between the whole market state, represented by ω, that is given by many, possibly interacting, random effects that we are not able to model, and are assumed to be random, and the results of our observation, i.e. the prices in the market that we want to model by our framework (Fig. 2.1).

In full generality, we give the following definition:

Definition.　Given a probability space $(\Omega, \mathcal{F}, \mathbb{P})$, a *random variable* is a real-valued function $X : \Omega \rightarrow \mathbb{R}$ such that for each measurable subset $B \in \mathbb{R}$, the subset of Ω given by

$$A = \{\omega \in \Omega | X(\omega) \in B\} \tag{2.4}$$

is in $\mathcal{F}$.

In this definition, we want to be sure that we can go back from each observation in the $\mathbb{R}$ space to the space of events Ω, where we defined a probability measure $\mathbb{P}$ (Fig. 2.2). In this way, whenever we want to measure

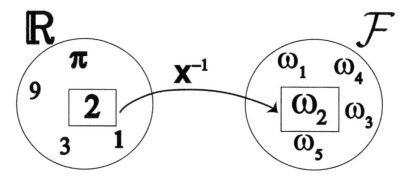

Fig. 2.2. Going back from each observation in the $\mathbb{R}$ space to the space of events Ω, we can measure the probability of the event through the probability measure $\mathbb{P}$.

the probability of $X(\omega)$, we are actually measuring the preimage of X into the Ω space.

Coherently, we introduce the following notation:

$$\mathbb{P}(X(\omega) \in B) = \mathbb{P}(\omega|X(\omega) \in B), \qquad (2.5)$$

where $B \subseteq \mathbb{R}$.

Under mild conditions on X, it can be proved that this approach actually defines a proper measure on $\mathbb{R}$ space that is called *push-forward measure* and is usually denoted by $\mu_X(x)$. We observe that this new measure μ does not explicitly depend on ω and, as a consequence, it can be directly inferred by empirical observations $X_i(\omega)$, where $i = 1, \ldots, n$ and n is the total number of observations. On the other side, the knowledge of μ_X does not imply the complete knowledge of $\mathbb{P}$; exactly in the same way, the knowledge about a price of the financial asset would not imply the knowledge of the state of the whole economic system. In any case, the push-forward measure $\mu_X(x)$ is fairly enough to deduce information and estimate $X(\omega)$, and it is what is generally considered for modeling purposes.

Once the push-forward measure is defined, one can exploit its definition in order to calculate the probability that $X(\omega) \in B$, extending Eq. (2.3):

$$\mathbb{P}(X(\omega) \in B) = \int_{X^{-1}(B)} \mathrm{d}\mathbb{P}(\omega) = \int_B \mathrm{d}\mu_X(x), \qquad (2.6)$$

where we stress that the first integral is defined in the Ω domain while the second one is defined in the $\mathbb{R}$ domain. If one interprets $\mu_{X(\omega)}(x)$ as a generic function $F(x) : \mathbb{R} \to [0, 1]$ and x as a dummy variable, one can think at the last integral in Eq. (2.6) as a standard Lebesgue–Stieltjes integral that can be solved by usual integral calculus. It follows that $\mu_{X(\omega)}(x) \equiv F(x)$ is a function that describes the probability of getting a random variable $X(\omega) < x$ as given by the following equation:

$$\mathbb{P}(X(\omega) < x) = \int_{-\infty}^{x} dF(u), \qquad (2.7)$$

where the function $F(x)$ represents the *cumulative density function* (CDF) of the random variable $X(\omega)$. In the same spirit, if $F(x)$ is a differentiable function, one can define the *probability density function* (PDF) $p(u)$ as

$$\int_{-\infty}^{x} dF(u) = \int_{-\infty}^{x} p(u)du. \qquad (2.8)$$

In practical applications, one tries to model the distribution of the random variables with different expressions of the CDF or the PDF in order to perform calculations in the usual integral framework. In this respect, we observe that in practice, it is not possible to access directly on Ω, so the CDF represents the whole available information on X. In Section 2.3, we will provide some examples of typical PDFs used in the financial context.

2.2.2. *The Information Flow Through Time: The Role of the Filtration*

In general, in modeling of activities and, in particular, in order to solve typical QF problems, it is necessary to formulate a mathematical description of the dynamics, i.e. the evolution through time of a generic observable of a system. In this respect, if a probabilistic approach is required, one needs to introduce the time parameter t in the previously described mathematical framework (Section 2.2). This operation is not theoretically straightforward as in general one should keep in mind that, as time passes, new sets of information are available, and they could potentially change the probability of future events. For example, we could think that at the very beginning of a soccer match, the two teams would have the same probability to win the match, i.e. 50%. If, after 85 mins, we knew that the first team won 10 goals to 0, we would say that the probability that the first team could win the match is almost 100%. In other words, our estimate of the probability has changed as a new information about our system was added. If we think a little bit more on the previous example, we would conclude that our estimation of the probability changed not because we changed the way we measure the amplitude of the set, but because actually the set corresponding to the final result 10−0 enlarged, increasing its volume (Fig. 2.3).

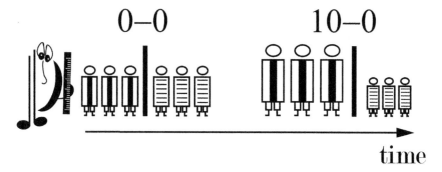

Fig. 2.3.   We show how the probability measure changes as new information is available. At the beginning of the match, the two teams seem to be of the same size, i.e. they have the same probability of winning the match. When the information about the partial result (10−0) is available, the team on the left is much larger.

This fact suggests that, in order to take into consideration the time evolution of a system and the information flow through time, one should not modify the probability measure but the σ-algebra over which the probability function acts.

In this spirit, we obtain the following definition:

Definition. We consider a non-empty set $\Omega \neq \emptyset$, $T \in \mathbb{R}^+$, and we assume that $\forall t \in [0, T]$ exists a σ-algebra $\mathcal{F}(t)$. In addition, we assume that $\forall s \leq t$, $\mathcal{F}(s) \subseteq \mathcal{F}(t)$. We define *filtration* as the collection of σ-algebras $\mathcal{F}(t)$, $0 \leq t \leq T$.

By filtration, we can model the information flow without a need to change the probability measure definition. The specific role of the filtration in the probability calculus is discussed in Section 2.3.

As a final remark, we observe that as the estimation of random variables could typically depend on the set of available information (represented by the filtration), this fact can be effectively exploited in order to define the concept of independence between random variables. Informally speaking, if the estimation of a random variable does not depend on the information in a σ-algebra $\mathcal{F}$, we can say that the variable is independent of $\mathcal{F}$. More formally, we can say that, given a probability space $(\Omega, \mathcal{F}, \mathbb{P})$, the two sub-$\sigma$-algebras $\mathcal{G}, \mathcal{H}$ (i.e. σ-algebras both included in $\mathcal{F}$) are independent if

$$\mathbb{P}(A \cup B) = \mathbb{P}(A)\mathbb{P}(B) \quad \forall A \subseteq \mathcal{G}, \forall B \subseteq \mathcal{H}. \tag{2.9}$$

Now, if X, Y are two random variables on $(\Omega, \mathcal{F}, \mathbb{P})$, we can say that X and Y are independent if the two σ-algebras that contain the information to estimate X and Y are independent. The notion of independence is of fundamental importance and it will be exploited in many cases in the following chapters for financial modeling purposes. In the following, we will extend our definition of probability space to $(\Omega, \mathcal{F}, \mathcal{F}(t), \mathbb{P})$ whenever we want to specify that the probability space is equipped by the filtration $\mathcal{F}(t)$.

2.3. How to Estimate a Random Variable — Expectations and Conditioning

In the preceding sections, we introduced the concept of randomness, and we provided a useful framework to describe it from a fundamental point of view. Once the randomness is included in a system, a very natural question arises if it is still possible to obtain a reliable *estimation* of the variables of interest, at least in a statistical sense. After all, in our framework, we introduced explicit probability *laws* (CDF) that characterize in an exact way how the randomness is included in the system we want to describe. As a consequence, even if it is difficult to deal with the value of a variable in usual deterministic way, one could expect to obtain, at least an *estimation* of the random value, taking also into consideration the information available at the time we want to obtain our estimation.

In this respect, a very simple idea could be to *weight* all the possible values of the random variable by the probability of obtaining each of them. Unfortunately, while dealing with variables defined on a continuous domain, the probability of getting each value is always zero,[a] and we are forced to talk about the probability that the random variable falls inside a finite interval. As a consequence, in order to weight each variable by a probability estimation, we need to express the weight in differential terms. We can then define the estimation of a random variable for a diffuse measure as

$$\mathbb{E}^{\mathbb{P}}(X) \equiv \int_{\Omega} X(\omega) d\mathbb{P}(\omega), \tag{2.10}$$

where we are weighting $X(\omega)$ by the probability of getting ω in a small interval $d\mathbb{P}(\omega)$. Sometimes, with a little abuse of notation, we refer to the

---

[a]We have already observed in Section 2.2 that a measure that assigns zero probability to the event of obtaining exactly a certain value is said to be a *diffuse* measure.

probability of getting $X(\omega)$ as

$$d\mathbb{P}(X(\omega)) \equiv \mathbb{P}\left(\omega | X(\omega) \in [x(\omega), x(\omega) + dx]\right) \tag{2.11}$$

and we equivalently define

$$\mathbb{E}^{\mathbb{P}}(X) \equiv \int_{\Omega} X(\omega) d\mathbb{P}(X(\omega)). \tag{2.12}$$

In Eqs. (2.10) and (2.12), we are assuming that for $X(\omega)$, there exists a probability space $(\Omega, \mathcal{F}, \mathbb{P})$, and that the (Lebesgue) integral is taken on the whole space Ω where the source of randomness ω can span. The term $\mathbb{E}^{\mathbb{P}}(X)$ is called *expected value* of the random variable X, and the operator $\mathbb{E}$ is called *expectation* operator. We observe that definition (2.10) depends on how we define the measure $\mathbb{P}$ and, in particular, two different measures of the same random variable could give different estimations; this is quite natural after all if we recollect that probability is simply the way we measure how likely an event could occur. In the rest of the book, we will neglect the probability specification in expectation notation whenever the probability measure is clear from the context.

This estimation of the random variable looks very similar to the usual definition of the sample mean (see Eq. (2.42)),

$$\hat{x} = \frac{1}{n} \sum_{i=1}^{n} x_i, \tag{2.13}$$

where n is the number of observations of the random variable. The expected value and the sample mean look very similar with the exceptions that (1) in the expected value definition, we are considering an integration instead of a simple sum as a consequence of the fact that we are dealing with continuous random variables and (2) the constant weight $1/N$ of Eq. (2.42) is substituted by the weight $d\mathbb{P}(X(\omega))$ that we know is normalized to one, as required by the first condition of Eq. (2.2). Actually, it will be shown in Section 2.6 that the sample mean estimator converges to the expected value when n becomes large as a confirmation that the two definitions are quite similar. In any case, we want to stress that the two definitions are deeply different from a theoretical point of view as the expected value calculation requires the knowledge of the probability measure that represents the most complete information we can have about the random variable. On the contrary, the sample mean is simply an estimator, where the information, that is contained in it, is based on a given number n of observations (*realizations*).

As already mentioned in Section 2.2.1, in practical applications, a complete knowledge of the probability measure $\mathbb{P}$ is hard to be achieved, and typically one relies on the push-forward measure CDF $F(x)$ or PDF $p(x)$ in order to perform calculations.[b] In this case, expectation definition can be restated as

$$\mathbb{E}^{\mathbb{P}}(X) = \int_{\mathbb{R}} x \mathrm{d}F(x) = \int_{\mathbb{R}} x p(x) \mathrm{d}x, \tag{2.14}$$

where the last equation on the right-hand side holds only if it is possible to define a proper PDF. By Eq. (2.14), one can obtain an estimation of the random variable X, just computing the integral with usual integration technique whenever the PDF is given. In Section 2.7, we will provide an example of this computation result for standard PDFs.

Definition of expectation given by Eq. (2.10) is essentially based on the knowledge of the whole probability measure $\mathbb{P}$, i.e. on all the available knowledge about randomness ω in our system. On the contrary, Eq. (2.14) is a little less demanding, and it requires only the knowledge of $F(x)$ that is a description of how the function $X(\omega)$ maps the information in ω into $\mathbb{R}$ (see the discussion on the measure push-forward in Section 2.2.1 and, in particular, the definition of random variable). As a consequence,

- Our knowledge of the whole randomness in Ω depends on how faithfully the information in ω is represented by $X(\omega)$.
- In order to be able to make an estimation of $X(\omega)$, we do not need all the information in $\mathbb{P}$, but only the information in $F(x)$ is enough.

These observations are of fundamental importance if one wants to take into consideration the information flow in defining a suitable estimation operator for the random variable $X(\omega)$. In other words, one could be interested in obtaining an estimation of a random variable that can be improved as new information is available. This fact is strongly related with what we already discussed in Section 2.2.2 where we introduced the concept of filtration as a way to model the information as the time passes by. Here, we want to obtain a definition of the expectation that depends on the information available at time t, i.e. on the σ-algebra $\mathcal{F}_t$.

---

[b]In this case, x is a dummy variable of the function $F(x)$ or $p(x)$. Sometimes, in order to remember that these functions refer to the distribution of the random variable $X(\omega)$, we add a subscript to the functions, $F_X(x)$ and $p_X(x)$.

Before proceeding with a formal definition, let us fix the ideas with a simple example. We consider an experiment made by an infinite sequence[c] of coin tosses $\omega = \{H, T, H, H, \dots\}$ (where H, T stands for *head* or *tail*, respectively), and we repeat this experiment infinite number of times, obtaining $\Omega = \{\omega_1, \omega_2, \dots\}$. We now define the random variable $X(\alpha, \beta)$ that depends only on the first two coin tosses, represented here by α and β. This implies that we do not need to wait an infinite number of extractions in order to know what is the exact value of X; a limited knowledge of the random space Ω is enough to obtain a good estimation of X and, in particular, just after two coin tosses, the random variable becomes *exactly known*. If there is enough information in the set (α, β) to determine the value of a random variable X, we say that X is (α, β)-*measurable*. Now, what we need to do is to include this aspect in our expectation definition, or in other words, we need to obtain a definition of the expectation that could update itself whenever a new information is available, improving the estimation of X.

For example, we could consider the situation after the first coin toss, i.e. when α is known. In this case, one could estimate the expected value taking into consideration the *condition* that $\alpha = H$ or $\alpha = T$:

$$\mathbb{E}(X | \alpha = H) = p_H X(H, H) + p_T X(H, T),$$

$$\mathbb{E}(X | \alpha = T) = p_H X(T, H) + p_T X(T, T), \tag{2.15}$$

where p_H and p_T are the probabilities of getting head or tail, respectively (equal to 0.5 if the coin is not biased), and the notation $\mathbb{E}(X | y)$ refers to the fact that the expectation is computed *assuming* that the condition y (represented by the σ-algebra) is verified or equivalently assuming that the information implied by y is available when we estimate the expectation. In fact, on the right-hand side of the equation, we considered only some possible values of the random variable (the ones with $\alpha = H$ in the first line and the ones with $\alpha = T$ in the second line); this is equivalent to assume that the other values have zero probability because the information expressed by y prevents it. In the following, we will refer to the notation $\mathbb{E}(X | y)$ as *conditional expectation*.

Unfortunately, the definition above is not well suited for continuous random variables, as the probability of a single event is zero, so an extension is required to the integral form. In order to do this, we first need to generalize Eq. (2.15) for a generic condition, represented by the σ-algebra $\mathcal{F}_t$.

---

[c]The fact that we require infinite coin tosses for each ω is because we want to deal with continuous random variables.

We define

$$\mathbb{E}\left(X|\mathcal{F}_t\right)\mathbb{P}(A) = \sum_{\omega \in A} X(\omega)\mathbb{P}(\omega) \quad \forall A \in \mathcal{F}_t, \tag{2.16}$$

where the condition $\omega \in A$ forces the sum to select only the events that belong to $\mathcal{F}_t$. In addition, we observe that in Eq. (2.15), we weighted the random variable X by the probability of the second coin toss, i.e. we considered what would happen *after* the condition. On the contrary, on the right-hand side of Eq. (2.16), we are weighting by the probability of getting the whole ω that also includes the probability of getting A; so, in order to keep the two equations consistent, we multiply the left-hand side of Eq. (2.16) by the probability of getting A.

We can now extend Eq. (2.16) to the integral representation, obtaining the standard definition of conditional expectation,

$$\int_A \mathbb{E}^{\mathbb{P}}\left(X|\mathcal{F}_t\right)(\omega)\mathrm{d}\mathbb{P}(\omega) = \int_A X(\omega)\mathrm{d}\mathbb{P}(\omega), \tag{2.17}$$

where we require that Eq. (2.17) holds $\forall A \in \mathcal{F}_t$.

We can observe that if $A = \Omega$, $\mathbb{P}(A) = 1$ by probability measure definition, and the conditional expectation degenerates into the expectation defined by Eq. (2.10).

As a consequence, Eq. (2.17) describes the generic operator $\mathbb{E}(\cdot|y)$ as a random variable that has the property of behaving as an expectation operator for every subset of Ω. In general, as suggested in Ref. [1], we refer to this aspect as *partial averaging* property that characterizes the conditional expectation. We want to stress that from this perspective, the conditional expectation is not only an estimation of a random variable given the information A, but it can also be understood as a random variable itself with the special property given by Eq. (2.17). In general, as the conditional expectation depends on the information in A, the random variable becomes completely known whenever the needed information is available, exactly as in the previous example, when, after two coin tosses, we were able to estimate the random variable $X(\alpha, \beta)$. Coherently with what we stated above, given a generic filtration $\mathcal{F}$, we say that X is $\mathcal{F}_t$-*measurable* if the information in $\mathcal{F}_t$ is enough to estimate the random variable X.

Starting from the definition of the expectation and conditional expectation, one can obtain useful relations with respect to the probability measure.

As an example, one could observe that the probability measure can be defined as the expectation of the indicator function $\mathbb{I}_x$ that is equal to one if the condition x is true and it is equal to 0 otherwise:

$$\mathbb{P}(x \in B) = \int_B d\mathbb{P}(\omega)$$

$$= \int_\Omega \mathbb{I}_{x \in B} d\mathbb{P}(\omega)$$

$$= \mathbb{E}(\mathbb{I}_{x \in B}) \tag{2.18}$$

$\forall B \in \mathcal{F}$. As a consequence, one can define the conditional probability relying on Eq. (2.17) as

$$\int_A \mathbb{P}(x \in B|A) d\mathbb{P}(\omega) = \int_A \mathbb{E}(\mathbb{I}_{x \in B}|\mathcal{F}) d\mathbb{P}(\omega)$$

$$= \int_A \mathbb{I}_{x \in B} d\mathbb{P}(\omega)$$

$$= \mathbb{P}(B \cap A). \tag{2.19}$$

Equation (2.19) can be understood as a formal definition of conditional probability that exploits the above-described measure theory framework and it is in good agreement with the classical definition of conditional probability that will be discussed in Section 2.4.

According to the notion of independence already discussed in Section 2.2.2, a random variable is independent on a σ-algebra if the information contained in it does not improve our knowledge of the variable. Coherently, one can prove that, if X is independent on the σ-algebra $\mathcal{F}$, the expectation and the conditional expectation are equal, i.e. the information in $\mathcal{F}$ does not improve our estimation of X:

• **Independence**

$$\mathbb{E}(X|\mathcal{G}) = \mathbb{E}(X). \tag{2.20}$$

In particular, exploiting Eq. (2.19), one can obtain that

$$\mathbb{P}(x \in B|\mathcal{F}) = \mathbb{P}(x \in B). \tag{2.21}$$

In order to conclude this section, we report, without demonstration, the main properties of the expectation operator in the case of conditioning, taking into consideration that they also hold for unconditional expectation. In the following, X, Y are two random variables, $\mathcal{H} \subseteq \mathcal{F}$ are two σ-algebras, α, β are two constants and $\phi(x)$ is a convex function of a dummy variable x:

- **Linearity**

$$\mathbb{E}(\alpha X + \beta Y | \mathcal{F}) = \alpha \mathbb{E}(X) + \beta \mathbb{E}(Y). \tag{2.22}$$

- **Iterated Conditioning**

$$\mathbb{E}(\mathbb{E}(X | \mathcal{F}) | \mathcal{H}) = \mathbb{E}(X | \mathcal{H}). \tag{2.23}$$

- **Jensen's Inequality**

$$\mathbb{E}(\phi(X) | \mathcal{F}) \geq \phi(\mathbb{E}(X | \mathcal{F})). \tag{2.24}$$

In addition, if X is completely determined by the information in $\mathcal{F}$, we can take it out of the conditional expectation:

- **Taking Out**

$$\mathbb{E}(XY | \mathcal{F}) = X \mathbb{E}(Y | \mathcal{F}). \tag{2.25}$$

2.4. Main Features of Probability Measure

In the preceding sections, we referred to probability concept in a very general (and perhaps too abstract) framework. The main risk related to this approach is to lose ourselves in technical details, missing simple aspects of the probability framework that are of fundamental importance in practical applications. In this section, we want to summarize the main features of the probability measure that can be used when calculations are performed in order to solve practical problems.

The first probability feature we want to stress was already introduced by Eq. (2.2) referring to the union of *disjoint sets*. In this case, we have (considering a finite and countable set of sub-samples)

$$\mathbb{P}(E_1 \cup E_2, \ldots, E_n) = \sum_{i=1}^{n} \mathbb{P}(E_i), \tag{2.26}$$

where $\{E_1, \ldots, E_n\}$ is the set of sub-samples. In other words, this equation tells us that the probability of getting one event *or* another one is exactly the sum of the probabilities of the two events *if* the events refer to different situations and, in particular, it is not possible that for a certain random configuration, the two events happen at the same time (*mutually exclusive events*). On the contrary, if the sets are not disjoint, Eq. (2.26) for two sets A, B becomes

$$\mathbb{P}(A \cup B) = \mathbb{P}(A) + \mathbb{P}(B) - \mathbb{P}(A \cap B), \tag{2.27}$$

where the notation $(A \cap B)$ refers to the intersection set. What is truly important to be kept in mind in practical applications is that the typical use of Eq. (2.26) is allowed only if events are disjoint; otherwise, a modification of the standard formula is required.

The second feature we want to discuss here refers to the concept of conditional probability which we have already discussed in Section 2.2.2. In this case, we expressed the conditioning as a filtration representing all the information available when the probability has to be estimated. Here, we want to refer to a more "classical" case of probability framework[d] where the notion of *available information* is simply expressed by the event we assume to happen for sure and, for this reason, it represents all the available information. Conditional probability answers the question: what is the probability of getting an event A *assuming* that an event B happens. The word "assuming" means that we are imposing a condition to the event A, requiring that another event B happens. This could remarkably influence the results of the probability calculation. For example, the probability of winning the lottery is typically very low, but if we condition the calculation to the assumption that we bought all the tickets, the probability of winning becomes equal to one. The conditional probability in the classical framework is defined as

$$\mathbb{P}(A|B) \equiv \frac{\mathbb{P}(A \cap B)}{\mathbb{P}(B)}, \tag{2.28}$$

where the normalizing factor $\mathbb{P}(B)$ has the role of assuring that event B surely happens for the calculation purpose. We observe that, as expected, Eq. (2.28) is in agreement with Eq. (2.19). In addition, from Eq. (2.21), it follows that if the events A and B are independent, whether the event B happens or not, it does not influence A probability calculation, as a consequence,

$$\mathbb{P}(A \cap B) = \mathbb{P}(A|B)\mathbb{P}(B)$$

$$= \mathbb{P}(A)\mathbb{P}(B). \tag{2.29}$$

Starting from this definition of conditional probability, one can obtain a relation between conditional and unconditional probabilities by simply conditioning on all possible events of the probability space and summing up all the obtained probabilities. In formal terms, given the set of disjoint subsets $\{B_1, \ldots, B_n\}$ such that $\Omega = \sum_{i=1}^{n} B_i$, where Ω is the whole space of

---

[d]We refer to classical probability framework as the probability theory formalized without the reference to measure theory and its axiomatic structure.

events, one has

$$\mathbb{P}(A) = \sum_{i=1}^{n} \mathbb{P}(A|B_i)P(B_i). \tag{2.30}$$

In general, we refer to Eq. (2.30) as *Law of Total Probability*.

The last feature of the probability we want to discuss in this section refers to the *Bayes's theorem*. This theorem provides a relation between the unconditional and conditional probabilities in the following sense:

$$\mathbb{P}(A_i|E) = \frac{\mathbb{P}(E|A_i)P(A_i)}{P(E)} = \frac{\mathbb{P}(E|A_i)\mathbb{P}(A_i)}{\sum_{j=1}^{n} \mathbb{P}(E|A_j)\mathbb{P}(A_j)}, \tag{2.31}$$

where $\{A_1, \ldots, A_n\}$ is a generic partition of Ω made by disjoint sets such that $\sum_{i=1}^{n} A_i = \Omega$. In the last passage of Eq. (2.31) we made use of Eq. (2.30). The demonstration of this theorem is quite simple and it is based on the definition of conditional probability (Eq. (2.28)):

$$\mathbb{P}(A|B)\mathbb{P}(B) = \mathbb{P}(A \cap B) = \mathbb{P}(B|A)\mathbb{P}(A). \tag{2.32}$$

Bayes's theorem is of fundamental importance in statistical applications because it describes how the *prior* probability estimation of an event $\mathbb{P}(A_i)$ is modified by the information in E, thus obtaining the *posterior* probability $\mathbb{P}(A_i|E)$. This theorem finds its practical application when one wants to improve the probability estimation of an event as the time passes and the information flows are updated.

2.5.   Moments and Cumulants

In the previous section, we showed how to obtain an estimation of a random variable and how to improve it taking into consideration the new information as the time passes by. We demonstrated that a good estimation can be obtained considering a weighted integral on all the possible values of the random variable $X(\omega)$, considering as a weight the probability measure $\mathbb{P}(\omega)$ in its differential form $d\mathbb{P}(\omega)$ (Eq. (2.10)). This equation can be interpreted in a different way, considering $X(\omega)$ as strange (diverging and not normalized) weight of the probability measure $\mathbb{P}(\omega)$. From this perspective, the solution of the integral represents a way to extract information about the probability measure $\mathbb{P}(\omega)$. In general, one could aim to increase the information on $\mathbb{P}(\omega)$, just repeating the integration for different functions of $X(\omega)$, each providing new additional information. For example, one could define

$$\mathbb{E}^{\mathbb{P}}(X^2) \equiv \int_{\Omega} X^2(\omega)\mathrm{d}\mathbb{P}(X(\omega)), \tag{2.33}$$

in order to obtain information about the *amplitude of the fluctuations* of the random variable $X(\omega)$, assuming that its expectation is equal to zero.

If we consider Eq. (2.33) in more detail, we can observe that $X(\omega)$ measures, in a quadratic sense, the distance of each variable from the zero expected value. From this intuitive explanation, it follows that the larger the magnitude of the fluctuations of X around its expected value, the larger the value given by Eq. (2.33). From this perspective, Eq. (2.33) provides an additional information about the probability measure $\mathbb{P}(\omega)$ as it implicitly describes what is the probability (in the case of symmetric probability measure) of getting an X-value close to 0: the smaller the value of Eq. (2.33) the higher the probability of getting X-values around 0. In this sense, Eq. (2.33) provides information about the amplitude of the distribution of $X(\omega)$, and it defines the so-called *second moment* of the random variable X.

Given Eq. (2.33), one could wonder what happens if the expectation of X is different from 0. In this case, one can generalize Eq. (2.33) and define

$$\mathrm{Var}(X) \equiv \mathbb{E}^{\mathbb{P}}\left(X^2\right) - \left(\mathbb{E}^{\mathbb{P}}\left(X\right)\right)^2, \tag{2.34}$$

where $\mathrm{Var}(X)$ is called the *variance* of the random process X. Typically, in practical applications, one is interested in an estimation of the fluctuations of X that has the same "unit of measure" of the expected value. For this reason, one considers

$$\mathrm{std}(X) \equiv \sqrt{\mathrm{Var}(X)}, \tag{2.35}$$

as a practical measure from the fluctuations amplitude. $\mathrm{std}(X)$ is called *standard deviation* of the random variable X.

In general, one can further extend the approach followed for second moment definition in order to get more information about the probability measure that refers to a generic random variable X. One can define

$$\mathbb{E}^{\mathbb{P}}(X^n) \equiv \int_{\Omega} X^n(\omega)\mathrm{d}\mathbb{P}(X(\omega)), \tag{2.36}$$

where n is an integer and $\mathbb{E}^{\mathbb{P}}(X^n)$ is called *nth moment* of the random variable X.

Moments definition can be obtained also considering the so-called *moment generating function*, defined as

$$M_X(t) = \mathbb{E}^{\mathbb{P}}(e^{tX}), \qquad (2.37)$$

where X is the random variable of interest and t is a dummy variable. It can be shown that the following relation holds between moments and the generating function:

$$m_n \equiv \mathbb{E}(X^n) = \left. \frac{\mathrm{d}^n M_X(t)}{\mathrm{d}t^n} \right|_{t=0}, \qquad (2.38)$$

so the knowledge of the moment generating function is enough to estimate all the moments of the random variable X. From another point of view, this fact seems to suggest that the knowledge of all the moments of a random variable provides quite a complete description of the moment generating function that actually is a generic transform of the probability measure.

Analogous to moments definition, one can introduce the so-called *cumulants* of a probability distribution, starting from the definition of the *characteristic function*,

$$h(t) = \mathbb{E}(e^{itX}), \qquad (2.39)$$

where i is the imaginary number and t is a dummy variable. If we look at Eq. (2.39) taking into consideration the integral definition of the expectation (Eq. (2.14)), we can observe that the characteristic function is closely related to the Fourier transform of the PDF of the random variable X except to the minus sign in the exponential term and the conventional factor $1/\sqrt{2\pi}$. Given the definition of the characteristic function, we can define the nth cumulant as

$$k_n = (-i)^n \left. \frac{\mathrm{d}^n \log(h(t))}{\mathrm{d}t^n} \right|_{t=0}. \qquad (2.40)$$

In general, there exists a close relation between moments defined by Eq. (2.38) and cumulants of Eq. (2.40); for the first four moments, the following relations hold:

$$\begin{aligned} m_1 &= k_1, \\ m_2 &= k_2 + k_1^2, \\ m_3 &= k_3 + 3k_2 k_1 + k_1^3, \\ m_4 &= k_4 + 4k_3 k_1 + 3k_2^2 + 6k_2 k_1^2 + k_1^4. \end{aligned} \qquad (2.41)$$

These relations can be extended to higher-order moments and cumulants as a confirmation that the two quantities refer essentially to the same information. In the following chapters of the book, we will alternatively make use of the notions of cumulants and moments as a way to characterize the PDF of a random variable.

2.6.   Statistical Estimators

Typically in statistics, one has to deal with two different aspects: On one side, there is the theoretical framework and theoretical distributions, that is, how we *think* random variables are distributed in theory; on the other side, we have the empirical data that represent the result of n-trials from an *unknown* distribution. What we want to do in statistics is to find a way to relate empirical observations and theoretical distributions. This is typically done by special functions, called *estimators*, of the realizations $\{x_1, \ldots, x_n\}$ that aim to estimate the underlying theoretical distribution or, at least, some of its properties. A typical example is given by the estimation of expectation by the sample mean of the realizations,

$$\hat{x} \equiv \frac{1}{n} \sum_{i=1}^{n} x_i. \tag{2.42}$$

From Eq. (2.42), it should be clear that the sample mean is itself a random variable as its value depends on the specific sample of random variables we use for the estimation. As a consequence, if we repeat many times the sample mean estimation on different samples, we would get results generally different from the expectation defined by Eq. (2.10). Given this situation, a very natural requirement is that, at least, the expectation of the estimator should be equal to expectation of the random variable,

$$\mathbb{E}(\hat{x}) = \mathbb{E}(x), \tag{2.43}$$

or in general, if θ is a generic parameter of the theoretical distribution $F(x; \theta)$, and $f(x_1, \ldots, x_n)$ is a θ estimator, we require that

$$\mathbb{E}(f(x_1, \ldots, x_n)) = \theta, \tag{2.44}$$

and we say that the estimator $f(\cdot)$ is *unbiased*. This property assures that if we try to estimate some theoretical value, starting from empirical data, we are sure that, at least, we get *on average* the right value. It will be shown in Section 2.8 that the sample mean is an unbiased estimator of the expected value of a distribution, as one could expect.

Analogously, it can be shown that

$$\hat{\sigma}^2 \equiv \frac{1}{n-1} \sum_{i=1}^{n} (x_i - \hat{x})^2, \tag{2.45}$$

is an unbiased estimator of the variance $\mathrm{Var}(X)$ of the random variable X.

In general, the property given by Eq. (2.44) is not assured to hold for every standard estimator and, in particular, it does not hold for typical estimators related to the extreme tails of the distribution (Section 2.7.4).

2.7. Probability Density Functions

In this section, we deal with common PDF and their analytical representation. As was already mentioned in Section 2.2, one can achieve a complete description of a random variable by the characterization of its push-forward measure $\mu_{X(\omega)}(x)$. In some cases, one can provide a representation of this measure by a density function whenever the following relation holds:

$$d\mu_{X(\omega)}(x) = p(x)dx, \tag{2.46}$$

where $p(x)$ is the PDF. It is interesting to observe that the $p(x)$-shape strongly depends on the system we want to describe and typically reflects the symmetries of the system itself. In particular, when the system has only one favorite state, the PDF is typically bell-shaped around this favorite state meaning that the configurations far from the typical configuration become less and less probable. In this respect, one can prove a very fundamental theorem which shows that the sum of many independent random variables (that satisfy some convergence requirements) converge in some sense to a bell-shaped distribution known as Gaussian distribution. The existence of this theorem, known as *Central Limit Theorem* (CLT) (see Section 2.8), induces in many cases the description of the PDF of interest by a Gaussian distribution. In this respect, we remark that, even if this assumption can represent a very good starting point for mathematical modeling, the convergence of this distribution is only assured when the number of observations is very large and for the central part of the distribution. On the other hand, what concerns the tails of the distribution, the Gaussian approximation could be very poor and not suitable for the modeling purposes. This aspect is especially relevant in the case of risk modeling, where the tails of the distribution are the most relevant part of the risk estimation.

In the following sections, we describe the main distributions mostly used in finance.

2.7.1. *Uniform Distribution*

In statistical modeling, a very natural assumption is to attribute to each event an equal probability; this assumption is quite recurrent when a "preferred value" for the system of interest is not evident, and potentially all values are equally probable. In these cases, one typically assumes that the random variable we want to model is *uniformly* distributed, and the PDF (in the continuous case) is given by

$$U(x; a, b) = \begin{cases} \dfrac{1}{b-a} & \text{for } a \leq x \leq b, \\ 0 & \text{for } x < a \quad \text{or} \quad x > b, \end{cases} \tag{2.47}$$

where $U(x; a, b)$ is the *Uniform* distribution, a, b are real numbers which describe the variation range of the random variable $X(\omega)$, and the denominator $b - a$ assures that the distribution is correctly normalized to one. The expectation and the variance of this distribution are given respectively by

$$\mathbb{E}(X) = \frac{1}{2}(a + b),$$

$$\text{Var}(X) = \frac{1}{12}(b - a)^2. \tag{2.48}$$

Despite its simplicity, the uniform distribution has a very fundamental role in statistical modeling as all random variables can be represented as a transformation of a uniform random variable by a suitable function. This property is formally stated by the following theorem: if X is a continuous random variable with cumulative distribution function F_X, then the random variable $y = F_X(x)$ has a uniform distribution on $U(x; 0, 1)$. As a consequence, if y has a uniform distribution $U(y; 0, 1)$ and F_X is a generic cumulative distribution, then the random variable $x = F_X^{-1}(y)$, obtained by the inverted function of $F(x)$, is distributed as F_X.

By this simple theorem, one can generate any random variable being able to generate uniformly distributed random variables and being able to calculate $F_X^{-1}(y)$. This theorem finds applications in the so-called Monte Carlo numerical simulations and in other fields of numerical analysis, where the generation of random samples with a given distribution is required.

2.7.2. *Bernoulli and Binomial Distributions*

The Bernoulli distribution describes the distribution of a random variable that has probability p of taking value 1 and $q \equiv 1 - p$ probability of taking value 0. Also in this case, one can imagine many situations where this distribution applies, especially in the so-called *hit-or-miss* experiment, where the value 0 or 1 represents the only two possible outcomes of an experiment. In this case, as the possible values of the random variable are discrete, it is more convenient to talk about probability mass function (PMF) instead of PDF

$$\text{Ber}(x; p) = \begin{cases} p & \text{for } x = 1, \\ 1 - p & \text{for } x = 0. \end{cases} \tag{2.49}$$

The expectation and the variance of this distribution are given respectively by

$$\mathbb{E}(X) = p,$$

$$\text{Var}(X) = p(1 - p). \tag{2.50}$$

In the same spirit, one could be interested in a more general framework, where one wants to estimate the probability of getting k-hits over n-shots, given that the probability of *a single hit* is given by p. In this case, the underlying distribution is *Binomial*, and the analytical expression of the PMF is

$$\text{Bin}(k, n, p) = \binom{n}{k} p^k (1 - p)^{n-k}, \tag{2.51}$$

where

$$\binom{n}{k} = \frac{n!}{k!(n-k)!} \tag{2.52}$$

is the binomial coefficient. We observe that, in the case $n = k = 1$, we recover the Bernoulli distribution. The expectation and the variance of the distribution are given respectively by

$$\mathbb{E}(X) = np,$$

$$\text{Var}(X) = np(1 - p). \tag{2.53}$$

Bernoulli and Binomial distributions have a wide range of applications and in particular can be useful to assess the efficiency of model forecasts (hits) with respect to the number of trials. We discuss this approach in Section 6.7 for the backtesting of risk models.

2.7.3. *Normal Distribution*

Normal (or Gaussian) distribution is probably the most famous statistical distribution and the most used distribution in mathematical modeling. Its fame is related to analytic tractability and its ability in approximating the asymptotic behavior of a large class of probability distributions, as stated by CLT. Here, it is discussed in Section 2.8. Sometimes, researches refer to this distribution as the *bell-shaped* because of its symmetric PDF centered around its expectation and exponentially decreasing behavior.

The analytic expression of the Gaussian PDF is given by

$$p_{\text{Normal}}(x; \mu, \sigma) = \frac{1}{\sqrt{2\pi\sigma^2}} \exp\left(-\frac{(x-\mu)^2}{2\sigma^2}\right), \tag{2.54}$$

where

$$\mathbb{E}(X) = \mu,$$

$$\text{Var}(X) = \sigma^2. \tag{2.55}$$

The CDF of the normal distribution is not explicitly known, but it can be expressed by a special function called *Error function*,

$$N(x; \mu, \sigma) = \int_{-\infty}^{x} \frac{1}{\sqrt{2\pi\sigma^2}} \exp\left(-\frac{(u-\mu)^2}{2\sigma^2}\right) du$$

$$= \frac{1}{2}\left(1 + \text{erf}\left(\frac{x-\mu}{\sigma\sqrt{2}}\right)\right), \tag{2.56}$$

where $N(x; \mu, \sigma)$ is the normal CDF and $\text{erf}(x)$ is the error function defined as

$$\text{erf}(x) = \frac{2}{\sqrt{\pi}} \int_0^x \exp\left(-u^2\right) du. \tag{2.57}$$

With a little abuse of notation, sometimes we will describe the Gaussian distribution of mean value μ and standard deviation σ by the notation $N(\mu, \sigma)$, neglecting the dummy variable x and without an explicit reference to the PDF or CDF form of the distribution. Gaussian properties are quite wide and they have large implications in many mathematical applications. We will discuss some of them in Section 2.8.

2.7.4. *Empirical Distribution*

In the following chapters, we show different models that aim to describe financial observables as random variables that belong to a postulated theoretical distribution. It is quite natural to expect that these models are accurate as much as the postulated distribution is in agreement with the actual distribution of these financial variables. At this point, a very natural question would be how to build an empirical distribution, i.e. a distribution based on empirical data, in order to compare our theoretical assumptions with what we actually observe in the market. The answer to this apparently simple question is not an easy task and should be considered with care. In particular, as it will be discussed in Chapter 6, the tail behavior is of fundamental importance for some financial applications as the risk estimation.

In this section, we discuss very simple approaches for PDF and CDF estimation referring to the specialized literature for more refined approaches.

Concerning the empirical PDF, the algorithm is quite straightforward and makes use of the definition of probability based on frequency. In particular, given a set of empirical realizations $\{x_1, \ldots, x_n\}$, one can split it in $m \ll n$ bins and count how many realizations fall inside each bin. A very natural choice is to linearly split the whole interval of constant length, i.e. $(x_n - x_1)/m$, but other choices can be considered in order to assure a larger number of realizations in the extreme bins. The assumption underlying this method is that the number of realizations that fall inside a bin is proportional to probability that a theoretical random variable with the same distribution of x belongs to the same bin. In general, in order to obtain a probability *density*, it is necessary to normalize the frequency count in order to assure that the integral over the whole x-range $[x_1, x_n]$ is equal to one, as required by probability definition. This task is typically performed considering standard numerical integration techniques.

Applying this simple method, one can obtain an empirical estimation of the PDF with the underlying assumption that *all the realizations in the sample are independent and belong to the same distribution* (Fig. 2.4). In addition, it should be observed that some arbitrariness is related to the function we consider for the bin split and the number of bins we want to consider. In particular, if m is comparable to the number of realizations n, the number of observations in each bin is small, and the statistical uncertainty increases significantly. As a consequence, the resulting empirical PDF shape will be very noisy. On the contrary, if one considers a small m, the resulting distribution could be too smooth hiding some relevant effects like peaks in the distribution (Fig. 2.5).

In practical situations, a reasonable compromise between m and n must be found empirically, considering the purpose and the focus of the analysis we need to perform.

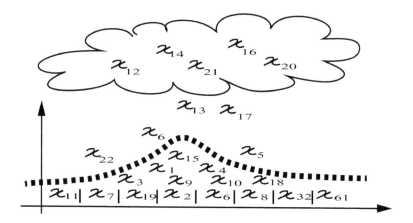

Fig. 2.4.   A representation of the empirical estimation of the PDF.

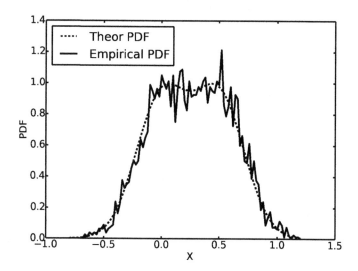

Fig. 2.5.   We compare the empirical PDF (solid curve) obtained considering 2000 realizations with its theoretical bi-modal distribution (dotted line). In this case, the number of observations is not large enough to clearly discern the two peaks of the theoretical distribution.

What concerns the empirical CDF, a very simple algorithm that is usually considered, is to order the sample of the realizations $\{x_1, \ldots, x_n\}$, and to associate to each ordered realizations x_i a probability equal to i/n, where i refers to the ordered position of x_i. In order to understand the rationale of this approach, we could consider, for example, the second ordered realization x_2 and its probability $2/n$. If our probability estimation is correct, it means that over n trials, we should expect $(2/n)n = 2$ realizations less or equal to x_2. Actually, this is exactly the case in our sample, as we have x_1 and x_2 that are by definition smaller or equal to x_2. As a consequence, $2/n$ should be a good candidate for the estimation of the CDF in x_2. Repeating this simple way of reasoning for each realization, we obtain a justification of the previously mentioned algorithm. Unfortunately, even if it looks reasonable, it can be shown that this approach is statistically meaningful only for the central part of the distribution, i.e. when $i \sim n/2$, and the related probability is about $i/n \sim 50\%$. On the contrary, the approach is less accurate in the tails of the distribution. This fact can be easily understood if we consider the maximum of the realizations x_n. In this case, we associate with this realization a probability equal to $n/n = 1$ that would imply that it is *never* possible to observe a realization larger than x_n even by repeating the extraction from the theoretical distribution an infinite number of times. This fact is intuitively wrong if the theoretical distribution is defined on a x-range that includes values larger than x_n (in the Gaussian case x goes to infinity!) and actually, by Extreme Value Theory, it can be proved that the probability of getting a realized value $x_{\hat{n}}$ larger than x_n considering another sample of $\hat{n}$-realizations is 67% for a wide class of distributions.

In general, we are dealing with the so-called problem of the *bias* estimator for a random variable (Section 2.6). In particular, it can be shown that if p is a given probability level and $x_{\text{th}}^p \equiv F^{-1}(p)$ is the value of the random variable x that assures that the probability of getting $x \leq x_{\text{th}}^p$ is p,[e] it can be shown that setting $i = np$, for $p \sim 0$ or $p \sim 1$,

$$\mathbb{E}(x_{\frac{i}{n}}) \neq x_{\text{th}}^p, \tag{2.58}$$

or in other words, the value of the ordered realization x_i is a bias estimator of the random variable x_{th}^p, when p refers to extreme events. On the contrary, it can be shown that x_i converges to the expected value x_p^{th} in the central part of the distribution, i.e. when the CLT holds, as will be discussed in Section 2.8.

In the following chapters, it will be shown that this approach in defining the empirical distribution is of fundamental importance in dealing with risk estimation. In this case, a special care should be taken in the estimation

---

[e]Here, the subscript "th" stands for theoretical value.

of events related to the tails of the distribution, especially the events that could negatively affect the value of financial portfolio.

To conclude this section, we show a nice trick to analyze the tail of an empirical distribution just plotting the empirical PDF on the semi-log scale (i.e. the y variable is substituted by $\log(y)$). In fact, given that a Gaussian distribution can be easily recognized as a parabola, taking the logarithm of Eq. (2.54), one can easily analyze the amplitude of the tails of the empirical distribution just comparing the empirical shape with a parabola. In particular, one refers to fat-tails distribution if the empirical PDF has a thinner body and larger tails.

As an example, in Fig. 2.6, we show the semi-log plot of the empirical distribution of price relative variations, compared to a Gaussian distribution $N(\mu_{\text{emp}}, \sigma_{\text{emp}})$, where we imposed μ_{emp} and σ_{emp} equal to sample mean and

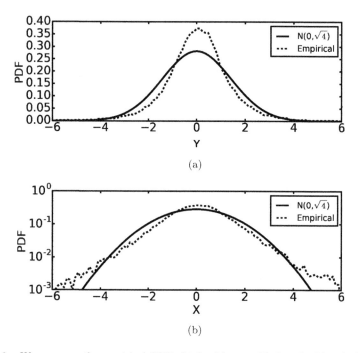

Fig. 2.6.   We compare the empirical PDF obtained by considering the historical series of price variations with a Gaussian distribution $N(\mu_{\text{emp}}, \sigma_{\text{emp}})$, where μ_{emp} and σ_{emp} are the sample mean and the sample variance, respectively. As expected, the semi-log plot magnifies the amplitude of the fat-tails of the empirical distribution with respect to the Gaussian one.

sample standard deviation, respectively. From the figure, it is evident that empirical price variations are not Gaussian and show fat-tails.

2.7.5. *Exponential Distribution and Student's t-Distribution*

As mentioned in the previous section, from a risk management perspective, what is truly relevant for risk estimation is the tail behavior of the distribution and, in particular, how it decays to 0 when x goes to $-\infty$ (or how it goes to 1 when x goes to $+\infty$). In this respect, a very important theory that refers to rare events is the Extreme Value Theory (EVT) Ref. [4] that derives asymptotic distributions of extreme events taking into consideration the distribution class of the random variables. Essentially, according to EVT, the CDFs of a random variable can be divided in to three classes:

- exponentially decaying distributions,
- power-law distributions,
- finite bound distributions,

where, for risk management purposes, most of the interest is in the first two classes. This simple classification of the CDFs somehow justifies our previous statement about the importance of the tail behavior in risk estimation. In particular, given EVT, one is allowed to focus on exponentially like distribution or power-law distribution in order to obtain reliable models for risk management purposes.

In this section, we provide two simple examples of distributions that belong to the above-mentioned distribution classes.

For what concerns the exponential tail class, we consider as representative distribution the so-called *exponential* distribution. The support of this distribution is given by $x \in [0, +\infty]$, and its PDF analytic expression is

$$p_{\exp}(x; \lambda) = \lambda \exp(-\lambda x), \tag{2.59}$$

where

$$\mathbb{E}(X) = \frac{1}{\lambda},$$

$$\text{Var}(X) = \frac{1}{\lambda^2}. \tag{2.60}$$

From the last expressions, we can observe that λ-parameter describes both the typical value (i.e. the expectation) and the typical fluctuations of the random variable. In addition, λ also characterizes the decay speed of the distribution, and for this reason, it is qualitatively related to the tail behavior of the distribution. The exponential CDF is given by

$$F(x; \lambda) = 1 - \exp(-\lambda x). \tag{2.61}$$

For what concerns the power-law decaying family, we consider, as an example, the so-called *Student's t-Distribution* (TS). In this case, the PDF is given by

$$p_{\text{TS}}(x; \nu) = \frac{\Gamma(\frac{\nu+1}{2})}{\sqrt{\nu\pi}\Gamma(\frac{\nu}{2})} \left(1 + \frac{x^2}{\nu}\right)^{-\frac{\nu+1}{2}}, \tag{2.62}$$

where ν is a parameter that represents the degrees of freedom of the distribution. Also in this case, the expectation and the variance are explicitly known:

$$\mathbb{E}(x) = \begin{cases} 0 & \text{for} \quad \nu > 1, \\ \text{undefined} & \text{for} \quad 0 < \nu \leq 1, \end{cases}$$

$$\text{Var}(x) = \begin{cases} \dfrac{\nu}{\nu - 2} & \text{for} \quad \nu > 2, \\ \infty & \text{for} \quad 1 < \nu \leq 2, \\ \text{undefined} & \text{for} \quad 0 < \nu \leq 1. \end{cases} \tag{2.63}$$

The Student's t-CDF is given by

$$F(x; \nu) = \frac{1}{2} + x\Gamma\left(\frac{\nu + 1}{2}\right) \frac{F_2^1\left(\frac{1}{2}, \frac{\nu+1}{2}, \frac{3}{2}, -\frac{x^2}{\nu}\right)}{\sqrt{\nu\pi}\Gamma(\frac{\nu}{2})}, \tag{2.64}$$

where $F_2^1(x)$ is a special function called *hypergeometric function*.

It can be shown that the TS distribution has a power-law decay,

$$p_{\text{TS}}(x; \nu) \sim \frac{\nu\Lambda_{\pm}^{\nu}}{|x|^{1+\nu}}, \tag{2.65}$$

where $\Lambda_{\pm}$ are two constants which describe the intensity of the decay for the negative and positive tails.

2.8.   Central Limit Theorem

As mentioned in Section 2.7.3, the Gaussian (or normal) distribution is probably the most famous (and the most used) distribution in practical applications. The reason for its "fame" is mainly due to the theorem we are going to discuss in this section, the so-called CLT. This theorem describes the behavior of the sum of a large number of independent and identically distributed (i.i.d.) random variables, and it demonstrates that *the central part* of the distribution converges in a distributional sense to a Gaussian distribution. Before discussing each detail of this theorem, we formalize it considering a set of n i.i.d. random variables $\{x_1, \ldots, x_n\}$ with expected value equal to μ, standard deviation equal to σ and defining $S(n) \equiv \sum_{i=1}^{n} x_i$. The theorem statement is as follows:

$$\lim_{n \to \infty} \mathbb{P}\left(u_1 \leq Z(n) = \frac{S(n) - \mu n}{\sigma \sqrt{n}} \leq u_2\right) = \int_{u_1}^{u_2} \frac{dx}{\sqrt{2\pi}} e^{-\frac{x^2}{2}}, \qquad (2.66)$$

where u_1 and u_2 are two *finite* real numbers. The statement of the theorem imposes some comments.

First, we need to explain the meaning of $\lim_n$. In mathematical terms, we refer to this kind of convergence as *convergence in distribution*, and this tells us that when n becomes very large, the *distribution* of $Z(n)$ cannot be distinguished from a Gaussian distribution with parameters $\overline{Z}(n) = 0$ and $\overline{Z^2}(n) = 1$. This does not imply that each realization of $Z(n)$ behaves as a Gaussian random variable in a one-to-one correspondence, but that from a distributional point of view, the two CDFs are very similar.

Second, we need to point out that the CLT holds only for finite u_1, u_2 and only when n goes to infinity. This implies that from an empirical point of view (i.e. when n is large but not infinite), we can expect that only the central part of the distribution of $Z(n)$ can be well approximated by a Gaussian distribution; on the contrary, the tails of the distribution cannot be described by the CLT. This fact is of fundamental importance when dealing with risk management aspects that typically refer to the tails of the distributions.

This effect is well represented in Fig. 2.7, where we show the distribution of the sum of n Student's t-random variables as n increases, and the corresponding CLT approximation of the distribution. As expected, as n increases, the central part of distribution is well approximated by Gaussian distribution, and it can be observed that the amplitude of this central region enlarges as n increases. On the contrary, the tails of the two distributions are not in good agreement, as the difference is magnified by the semi-log scale representation.

CLT can also be used in order to identify the statistical behavior of the sample mean estimator $\hat{x}$ as defined by Eq. (2.42). In this respect, as already mentioned in Section 2.6, we know that the sample mean is an unbiased statistical estimator of the expected value of random variables.

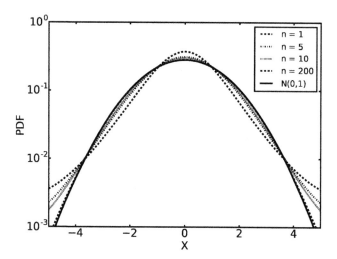

Fig. 2.7. We show the convolution of n Student's t-PDFs with $\nu = 4$ degrees of freedom (properly normalized by a factor $\sqrt{n}$) as n increases, and we compare them with a Gaussian distribution $N(0, 1)$. Because of the CLT, the convoluted distribution approaches the Gaussian one in the central region, while in the tails, the two distributions are still different. On the other side, the convergence region increases with n.

This fact is assured by a theorem called *Law of Large Numbers* (LLN). In addition, the CLT gives us an important information about *how* the sample mean converges to the expected value μ, providing a useful approximation of the central part of the distribution of $\hat{x}$. In formulas, the CLT can be restated as

$$\lim_{n \to \infty} \left| \mathbb{P} \left(u_1 \leq \hat{x}(n) \leq u_2 \right) - \int_{u_1}^{u_2} \frac{dx}{\sqrt{2\pi \left(\frac{\sigma^2}{n} \right)}} e^{-\frac{(x-\mu)^2}{2\left(\frac{\sigma^2}{n} \right)}} \right| = 0, \qquad (2.67)$$

considering a simple change of variables in Eq. (2.66). From Eq. (2.67), we can observe that the CLT confirms that the sample mean is an unbiased estimator of the expected value of x, as the Gaussian distribution is centered around μ. In addition, CLT provides us an estimation of the standard deviation of the estimator as $\sigma/\sqrt{n}$. This is a very useful result from a practical point of view because it gives us a very important information about how much our estimator will randomly fluctuate around the "correct" value μ. In this respect, the result is quite intuitive: the fluctuations of the sample mean depends on the amount of noise, σ, involved in the original random variable x, and *divided by* the square root of the number of (i.i.d.) observations which we used to estimate the sample mean. This justifies the use of the sample mean as an estimator of the expected value of a distribution. Also, the sample mean estimator is considered as a smoothing function for random variables, as it implies a reduction factor proportional to $1/\sqrt{n}$ of the original noise. In this respect, we observe that n represents the number of sources of independent information, and this is a crucial aspect in sample mean error estimation. In particular, if realizations are correlated, the noise reduction factor $1/\sqrt{n}$ can be much larger than the one which can be obtained by the same number of independent observations.

At this point, one could think that the CLT is just a nice way to approximate the expected value and the standard deviation of the sample mean. Actually, it should be pointed out that information included in the CLT is much more complete, as it provides also a careful description of the *probability* of the sample mean estimator. In fact, from the CLT, we can also estimate the probability of getting a sample mean value at a given distance from the expected value of the distribution, just by integrating the Gaussian PDF.

2.9. Stochastic Processes

In all theoretical frameworks presented above, we essentially were dealing with a single random variable of the corresponding distribution. In practical applications, this corresponds to observation of a random (stochastic) system at a fixed time, and analyzing the statistical properties of the variables of interest at the specific time. For example, in financial applications, this framework can be useful to describe what could be the distribution of the price of an asset over a given time horizon. In addition, we introduced in Section 2.2.2, an approach to take into consideration the information flow as the time passes by, considering the mathematical concept of filtration.

What is still missing is a tool to describe how the random variable of interest can *evolve* through time or, in other words, how we can describe the *dynamics* of the random variable. This is clearly a fundamental requirement for our theoretical framework, as typically one is interested not only in the distribution of the prices at a given time horizon but also in how the final price is reached. In fact, from a financial point of view, it is completely different if we can increase the value of our portfolio by, say, 20% in one year without any potential losses during this time, or if we are forced, in order to realize this gain, to expose our portfolio to large fluctuations during this year. So, in order to obtain a complete description of the performance of our investment, the final distribution of its prices is not enough, and the dynamics of the prices must be taken into consideration.

In order to overcome this deficiency in the theory, we need to introduce the concept of stochastic process that, in mathematical terms, is simply a *family of random variables labeled by the parameter t* that describes the time. In the rest of the book, we will refer to a stochastic process by the notation $X(t, \omega)$, $X_t(\omega)$, or simply $X(t)$, in order to describe this family of functions that changes stochastically as a consequence of the random variable ω.

In Fig. 2.8, we present different paths of a stochastic process $X(t, \omega)$ as a function of time. As evident from the figure, the paths in this case are all continuous, but their erratic behavior does not allow the definition of the derivative in each point. The vertical black lines represent the PDFs at different fixed times. The stochastic process, represented in the figure, is called Brownian motion, and it will be discussed in the next section.

In general, given that a random variable is nothing but a stochastic process with a fixed value of the parameter t, all definitions given above for the random variables also apply in the context of a stochastic process with the usual meaning. In particular, one can estimate the expectation and the conditional expectation of a stochastic process, using standard tools of the random variables calculus. In the following sections, we will focus on the

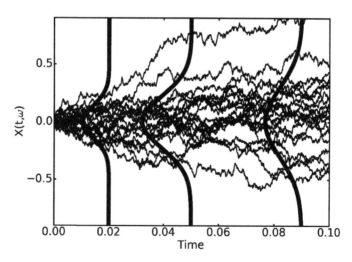

Fig. 2.8.   Different paths of a famous stochastic process, the Brownian motion. The bold lines represent the PDFs at different fixed times.

simplest example of the stochastic process, i.e. the Brownian motion, but the reader should keep in mind that it just represents one of the simplest dynamics that one could obtain in this framework.

2.10. Brownian Motion

In this section, we want to introduce the most famous and probably the most used example of stochastic process, the so-called Brownian motion. This process represents a very good starting point for the description of many physical and financial problems, and it will be used throughout the rest of the book in order to model the price evolution of an asset.

As we did in Section 2.3, when we introduced the idea of the conditional expectation, we first analyze here a discrete random process called *random walk* following the lines in Ref. [1] in order to become more familiar with the stochastic processes. Then, we will obtain the Brownian motion as a limiting case.

We consider an infinite sequence of (discrete) coin toss results, H or T, and we define

$$X_j \equiv 1 \quad \text{if } \omega_j = H,$$
$$X_j \equiv -1 \quad \text{if } \omega_j = T, \tag{2.68}$$

where j labels the jth coin-toss. In addition, we define $M_0 = 0$ and

$$M_k \equiv \sum_{j=1}^{k} X_j, \tag{2.69}$$

where $k = 1, 2, \ldots$ and M_k is our *symmetric random walk*. If we think of k as a discrete version of time, M_k can be understood as time-discrete stochastic process. We now choose a sequence of (not necessarily consecutive) ordered integers $0 = k_0, \ldots, k_m$, and we define the *increment*,

$$\Delta I_{k_i} \equiv M_{k_{i+1}} - M_{k_i} = \sum_{j=k_i+1}^{k_{i+1}} X_j. \tag{2.70}$$

The following properties hold:

- As X_j is independent, and the sum of independent random variables is an independent random variable, we conclude that ΔI_{k_i} is also an independent random variable.

- The expectation of X_j is 0, i.e.

$$\mathbb{E}(X_j) = 0. \qquad (2.71)$$

- The variance of X_j is equal to 1, i.e.

$$\text{Var}(X_j) = 1. \qquad (2.72)$$

- The variance of ΔI_{k_i} linearly increases with the time increment,

$$\text{Var}(\Delta I_{k_i}) = \sum_{j=k_i+1}^{k_{i+1}} \text{Var}(X_j) = \sum_{j=k_i+1}^{k_{i+1}} 1 = k_{i+1} - k_i. \qquad (2.73)$$

Now, we would like to obtain the limiting case by increasing the number of tosses in the unit of time t. Unfortunately, we cannot simply define M_{nt}, as we would have that $\text{Var}(M_{nt}) \to +\infty$ as $n \to +\infty$. In order to normalize the limiting case, we define the *symmetric scaled random walk*,

$$W^n(t) \equiv \frac{1}{\sqrt{n}} M_{nt}, \qquad (2.74)$$

so that we have (assuming $t > s$)

$$\mathbb{E}(W^n(t)) = 0,$$

$$\text{Var}(W^n(t)) = t,$$

$$\mathbb{E}(W^n(t) - W^n(s)) = 0, \qquad (2.75)$$

$$\text{Var}(W^n(t) - W^n(s)) = t - s,$$

$$\mathbb{E}(W^n(t)|\mathcal{F}(s)) = W^n(s).$$

In addition, we can observe that, as all random variables X_j are independent, the CLT holds (see Section 2.8), so we have that when $n \to +\infty$,

$$W^n(t) \to N(0, \sqrt{t}), \qquad (2.76)$$

where N is the CDF of a normal distribution with zero mean and variance equal to t.

We are now ready to define the Brownian motion as the limiting case $(n \to +\infty)$ of a symmetric scaled random walk.

Definition. We consider a probability space $(\Omega, \mathcal{F}, \mathbb{P})$, and for all $\omega \in \Omega$, we assume there exists a continuous function $W(t, \omega)$ $(t \geq 0)$ such that $W(0, \omega) = 0$ for all ω.

$W(t)$ is a *Brownian motion*, or a *Wiener process* if and only if for all sub-sequences $\{0 = t_0 < t_1 < \cdots < t_m\}$, all the increments $W(t_1) - W(t_0)$, $W(t_2) - W(t_1), \ldots, W(t_m) - W(t_{m-1})$ are independent and normally distributed such that

$$\mathbb{E}(W(t_{i+1}) - W(t_i)) = 0,$$

$$\text{Var}(W(t_{i+1}) - W(t_i)) = t_{i+1} - t_i. \tag{2.77}$$

The notation $W(t)$ is named in the honor of Norbert Wiener.

We observe that in the case of the symmetric random walk, we obtained the normal distribution as a limiting case when n goes to infinity. On the contrary, in the case of the Brownian motion, we *required* by definition that the increments are normally distributed, avoiding any approximations. In addition, we note that we required that the function $W(t)$ has to be continuous in time. This fact is of fundamental importance in financial models because it assures that strong disruptive events cannot affect the dynamics of the price we are going to model. This represents a very strong working hypothesis as, from an empirical point of view, quite often discontinuities can be observed in price dynamics. On the other side, the continuity of the path assures that, whatever is the noise we introduced into the process, each point of the process is still "connected to the previous one". This has an important theoretical implication, as it would suggest that even if we are not able to completely forecast the future dynamics of the price we want to model, we are, at least, in the position to optimally *adapt* ex-post our trading strategies, knowing that the future price cannot lie too far from the current position. From this intuition, it could be reasonable to expect that frequently adapting our decisions to the current market situations would lead to an optimal behavior of our investment strategy. This would represent an informal justification of the so-called *Martingale Representation Theorem* that will be stated in Section 2.15.1.

In general, the list of properties of the Brownian motion is quite long as well as the implication in applied stochastic modeling. In the following sections, we will discuss in more detail a few main properties that characterize it with no claim to be exhaustive. To conclude this section, we recall two Brownian motion basic properties

$$\mathbb{E}(W(t)) = 0,$$

$$\text{Var}(W(t)) = t. \tag{2.78}$$

2.11. Quadratic Variation

In the preceding sections, we introduced the concept of random variable, stochastic process and, more specifically, Brownian motion. The idea underlying this mathematical framework is that in our system, there is some noise that cannot be described in detail and it is required to be modeled as a random term. In this respect, a very natural question could be what are the mathematical implications of this choice and, in particular, what is the main difference between a deterministic function of time $f(t)$ and a stochastic function of time, i.e. a stochastic process, $X(t, \omega)$. At first sight, one could expect that, because of the presence of the random term, the stochastic process should look "more erratic/fluctuating" in some sense. On the other side, given the definition of Brownian motion (see Section 2.10), we already know that both functions can be continuous, so the fluctuations of the stochastic process should appear in a way that does not destroy the continuity of the path. In addition, as we obtained the Brownian motion as a limiting behavior of a random walk, requiring that the time unit becomes increasingly small, we should expect that the stochastic nature of the process cannot disappear below a given time unit.

Given this preliminary observations, we focus our interest on the nature of the fluctuations of a path on a very small unit of time, and we try to define a sort of time average of the fluctuations. In formal terms, we define the *Quadratic Variation* (QV) as

$$[X, X](T) \equiv \lim_{||\Pi|| \to 0} \sum_{j=0}^{n-1} (X(t_{j+1}) - X(t_j))^2, \qquad (2.79)$$

where $\Pi = \{x_0, \ldots, x_n\}$ is a generic partition of the interval $[0, T]$ and $||\Pi|| = \max_{1 \leq k \leq n}(x_k - x_{k-1})$ is a measure of the partition.

Equation (2.79) measures the amplitude of the variations of the stochastic process as the partition of the interval becomes small. It can be demonstrated that the QV of a function defined on the interval $[0, T]$ with a continuous time derivative is zero. This is intuitively true for the following reason: the time derivative of a function represents how much the function varies though time, so its squared value can be considered as an estimation of the amplitude of the fluctuations. As we know that the derivative is continuous, we cannot expect large variations on small time step. On the contrary, as the partition measure goes to zero, the time step over which we measure the variation goes to zero, and this dominates any possible divergence of the time derivative. As a consequence, the QV goes to zero whenever the time derivative is continuous.

On the other side, it can be proved that the quadratic variation of the Brownian motion is almost everywhere[f] equal to T, for all $T \geq 0$. In formulas, we write

$$[W, W](T) = T. \tag{2.80}$$

This very interesting result is strongly related to the fact that the variance of the Brownian motion scales linearly with time Eq. (2.78), and it suggests the true nature of this stochastic process. Qualitatively speaking, the main difference between a deterministic function and the Brownian motion function is that if we observe the first one on a very small time scale, we would obtain a very smooth function; on the contrary, a Brownian motion still presents fluctuations that cannot be avoided further decreasing the time unit. This is essentially due to the fact that the variation $W(t + dt) - W(t)$ is purely random; as a consequence, the presence of fluctuations is an intrinsic feature of the function $W(t)$ that cannot be avoided, decreasing the time scale.

It would be tempting to consider the QV equivalent to the variance defined by Eq. (2.34). Actually, this would be justified as the two objects give the same result in the case of the Brownian motion, and they seem to represent the same concept. This is typically exploited in practical applications when one has empirical data referring to a single path and wants to obtain an estimation of the variance of a process. In this respect, we want to stress that the variance is an average over the different paths of a stochastic process (we fix t and we integrate over ω), on the contrary, the QV is similar to a *time*-average on the same path (we fix ω and sum over t). This aspect can be of fundamental importance from the theoretical point of view, and it should be kept in mind in developing stochastic models.

2.12. Martingale

Naively speaking, one could expect that the main ingredient of the speculative finance is what we think is the future price of an asset. Unfortunately, price forecast is not an easy task, in most of the cases, and it cannot be

---

[f]By *almost everywhere*, we refer to a mathematical jargon used to say that even if there are paths that violate a given property, the probability of getting these paths is equal to zero.

achieved by simple statistical models. On the other side, in Section 3.1, it will be shown that models that can provide prices of complex financial instruments (pricing models) actually do not need this kind of forecasts as long as one is interested in the price of the instruments that can be considered *fair* for all market participants. The exact meaning of *fair value* will be clarified in the following chapters, when an appropriate theory will be developed. Here, we just want to show how the concept of fairness is related to a very important property of the stochastic process, that is, the so-called *martingale* property.

In formal terms, a stochastic process $X(t)$ is a martingale if

$$\mathbb{E}(X(t)|\mathcal{F}(s)) = X(s) \qquad \forall s,t \quad s \leq t \leq T, \tag{2.81}$$

where $\mathcal{F}(t)$ is a filtration on the probability space $(\Omega, \mathcal{F}, \mathbb{P})$ and T is a positive fixed number. Essentially, this relation tells us that the best estimation of the future value of $X(t)$, given its current value $X(s)$, is exactly $X(s)$. So, if we think at $X(t)$ as the price of a financial instrument at time t, the martingale property tells us that, on average, neither gains nor losses should be expected from an investment in this asset.

From this perspective, the concept of martingale is strongly related to the concept of *fair game*, as nobody can have any *priori* advantage by this investment.

It can be shown that the Brownian motion is a martingale, so the following equation holds:

$$\mathbb{E}(W(t)|\mathcal{F}(s)) = W(s) \qquad \forall s,t \quad s \leq t \leq T. \tag{2.82}$$

2.13. Stochastic Differential Equations (SDEs)

In Section 2.10, we introduced the Brownian motion as the most important stochastic process in QF. As a matter of fact, in the following chapters, we are going to exploit Brownian motion properties in order to model the *random part* of price dynamics. On the other side, it would be helpful to add some degree of freedom in our formalism in order to model contextually the deterministic part (if any) of the asset price dynamics and the noisy one. For this reason, we could try to express the dynamics of the asset price in differential terms, where the price variation can be described by a deterministic function *plus* a random part. In formulas, we have

$$dS(t,\omega) = \mu(S,t)dt + \sigma(S,t)dW(t,\omega), \tag{2.83}$$

where $S(t, \omega)$ is a random variable which represents the price, $\mu(S, t)$ is a function that describes the deterministic part of our equation, and it is called *drift*, and $\sigma(S, t)$ is a function that controls the amplitude of the noise and it is called *diffusive* term. By Eq. (2.83), we are quite free to model the price dynamics in the case of continuous random paths generated by $W(t)$, by specifying the functions μ and σ. Additional extensions of Eq. (2.83) can be achieved, for example, by the introduction of further random terms belonging to different distribution families. In the rest of the book, we will focus only on the Brownian-like processes.

In order to completely understand Eq. (2.83), we need to clarify the meaning of the differential notation in the stochastic framework. In this respect, we have already observed in Section 2.11 that the derivative definition for stochastic process could lead to pathological situations, as the QV does not go to zero when the time unit becomes infinitesimal. On the other side, in the same section, we observed that the QV is zero whenever the time derivative is continuous. As a consequence, we should expect difficulties in defining in a proper way the derivative of a stochastic process.

On the other side, even if a derivative definition is not available, it is still possible to properly define the integral of a stochastic process simply relying on the sum definition and considering its limiting case. In formulas, considering a time interval $[0, T]$, a partition $0 = t_0, t_1, \ldots, t_n = T$, and a generic function $f(t, \omega)$, we define the integral of the Brownian motion as

$$\int_0^T f(t, W(t)) \mathrm{d}W(t) \equiv \lim_{n \to \infty} \sum_{i=0}^{n-1} f(t^*(i), W(t^*(i))) \left(W(t_{i+1}, \omega)\right.$$
$$\left. - W(t_i, \omega)\right), \tag{2.84}$$

where $t^*(i)$ is a generic point in the interval $[t_i, t_{i+1}]$. This integral definition is similar to the standard Riemann integral definition with the exception that in this case, the differential part is given by the stochastic element $\mathrm{d}W(t)$. In this respect, we have already discussed in Section 2.11 the properties of the stochastic increment, and we observed that because of its randomness, it is not possible to know its value *before* the time t_i. For the same reason, in general, the value of the function $f(t^*(i)W(t^*(i))$ is not known for all $t^*(i) > t_i$. The only exception is represented by the value of $f(t_i, W(t_i))$ that is known at time t_i as it does not depend on $\mathrm{d}W(t)$. As a consequence, different from the Riemann integral, the result of the integral of Eq. (2.84) *depends* on the time $t^*(i)$ we choose to evaluate the function, and consequently, we can obtain different integral definitions for different choices of $t^*(i)$. Probably, the most famous choice is to take $t^*(i) = t(i)$, assuring that the function $f(t_i, \omega)$ is known at each time. This choice provides the definition of the *Itô's integral*,

$$\int_0^T f(t, W(t))\mathrm{d}W(t) \equiv \lim_{n \to \infty} \sum_{i=0}^{n-1} f(t_i, W(t_i))(W(t_{i+1}, \omega) - W(t_i)). \quad (2.85)$$

If we analyze in detail Eq. (2.85) and, in particular, the sum on the right-hand side, we can observe that each term of the sum is made by a product of a "fixed" part $f(t_i, W(t_i))$, known at $t = t_i$, multiplied by a stochastic increment with zero mean. For this reason, we will generally refer to the *non-anticipating* (or *adapted*) nature of the Itô's integral, as forecasts of the function $f(t, W)$ are not required in order to estimate the integral.

In general, we can define an *adapted* (*non-anticipating*) *process*, a stochastic process $X(t)$, that is, $\mathcal{F}(t)$-measurable for each $t \in [0, T]$.

Given this simple observation and exploiting the linearity of the expectation operator, we can obtain a standard result of the stochastic calculus,

$$\mathbb{E}\left(\int_0^T f(t, W(t))\mathrm{d}W(t)\right) = 0, \quad (2.86)$$

and in general, it can be shown that the Itô integral, as defined by Eq. (2.85) is a *martingale*.

Analogously, exploiting the fact that the QV of a Brownian motion scales with time, the so-called *Itô's Isometry* can be demonstrated [5]:

$$\mathbb{E}\left[\left(\int_0^T f(t, W(t))\mathrm{d}W(t)\right)^2\right] = \mathbb{E}\left[\int_0^T f^2(t, W(t))\mathrm{d}t\right]. \quad (2.87)$$

In other words, this result shows that it is possible to bring the square inside the stochastic integral, transforming the stochastic differential term $(\mathrm{d}W)^2$ into the time differential $\mathrm{d}t$.

As already mentioned above, this property is essentially due to a QV of the Brownian motion (Eq. (2.80)) that is usually written as

$$\mathrm{d}W(t)\mathrm{d}W(t) = \mathrm{d}t. \quad (2.88)$$

This notation is extremely powerful from a mnemonic point of view, as it tells us how to deal with the product of stochastic differentials. Unfortunately, Eq. (2.88) is simply wrong as we cannot be sure that the square of a Gaussian random numbers (with zero mean and variance $\mathrm{d}t$), i.e. $\mathrm{d}W(t)$, is equal to dt. As a consequence, Eq. (2.88) must be understood as a short way to refer to the integral expression,

$$\int_0^T dW(t)dW(t) = \int_0^T dt = T, \qquad (2.89)$$

that we know holds because of Eq. (2.80). This way to interpret the short-hand differential notation applies to all SDEs used in this book. In particular, we now clarify the meaning of Eq. (2.83) that must be interpreted as

$$\int_t^{t+dt} dS(t,\omega) = \int_t^{t+dt} \mu(S,t)dt + \int_t^{t+dt} \sigma(S,t)dW(t,\omega). \qquad (2.90)$$

Additional differential relations that can be useful for stochastic calculus purposes are

$$dtdt = 0,$$

$$dW(t)dt = 0. \qquad (2.91)$$

In this section, we clarified the meaning of the differential notation in a stochastic theoretical framework, avoiding the derivative definition and relying on the Itô's integral definition. In the next section, we are going to describe how to deal with SDE when functions of stochastic variables are involved, obtaining a suitable expansion in differential terms.

2.14. Itô's Lemma

In the previous section, we discussed on how to define SDE in order to model price dynamics. We observed that because of the random nature of the Brownian motion, its QV is not zero, and we cannot properly define the time derivative of the process. As a consequence, we had to work with differential terms, meaning that an integral properly defined is always implied.

At this point, a very natural question arises, if given the SDE for the process $X(t,\omega)$ and a smooth function $f(t, X(t,\omega))$, it is possible to obtain the SDE for $f(t, X(t,\omega))$. In other words, we are looking for the stochastic equivalent of the first-order chain rule for deterministic function $g(x,y)$,

$$dg(x,y) = \frac{\partial g(x,y)}{\partial x}dx + \frac{\partial g(x,y)}{\partial y}dy. \qquad (2.92)$$

In the stochastic framework, Eq. (2.92) is not valid anymore as, by Eq. (2.88), we know that also second-order dW terms contribute to the first-order time corrections dt. Actually, it can be proved that the stochastic equivalent of the chain rule can be obtained by also considering the differential QVs of the stochastic process. This result is known as *Itô's lemma* [5,6], and it is described by the following relation:

$$df(t, X(t,\omega)) = \frac{\partial f(t,x)}{\partial t}dt + \frac{\partial f(t,x)}{\partial x}dX(t,\omega)$$

$$+ \frac{1}{2}\frac{\partial^2 f(t,x)}{\partial x^2}d[X(t,\omega), X(t,\omega)], \qquad (2.93)$$

where d$[X(t,\omega), X(t,\omega)]$ is the QV of $X(t,\omega)$ expressed in differential terms. We observe that the partial derivatives in Eq. (2.93) must be understood as derivatives with respect to dummy variables t, x, forgetting that x represents a stochastic process. In order to stress this fact, we used the notation $\partial/\partial x$.

Equation (2.93) allows one to switch from the SDE to another one just giving the expression of $f(t, x)$ and calculating partial derivatives. In particular, assuming the SDE given by Eq. (2.83), one obtains that the SDE for the generic smooth function $f(t, X(t,\omega))$ is given by

$$df(t, X(t,\omega)) = \frac{\partial f(t,x)}{\partial t}dt + \mu(S,t)\frac{\partial f(t,x)}{\partial x}dt$$

$$+ \sigma(S,t)\frac{\partial f(t,x)}{\partial x}dW(t) + \frac{1}{2}\sigma^2(S,t)$$

$$\times \frac{\partial^2 f(t,x)}{\partial x^2}dt, \qquad (2.94)$$

where, in order to estimate the differential QV of $X(t,\omega)$, we exploited Eq. (2.88).

The Itô's lemma represents one of the building blocks of the stochastic calculus, and it will be used in the following chapters in order to model asset price dynamics.

As an example, one can use the Itô's lemma to solve a particular case of Eq. (2.83), when the drift and the diffusion coefficients, $\mu(S,t)$ and $\sigma(S,t)$, are proportional to $S(t)$, i.e.

$$dS(t,\omega) = \mu(t)S(t)dt + \sigma(t)S(t)dW(t,\omega). \qquad (2.95)$$

In this case, one can apply the Itô's lemma to the function $H(S(t)) = \log(S(t))$,

$$dH(t) = \frac{1}{S(t)}dS(t) - \frac{1}{2}\frac{1}{S^2(t)}\sigma^2(t)S^2(t)dt$$

$$= \left(\mu(t) - \frac{\sigma^2(t)}{2}\right)dt + \sigma(t)dW(t), \qquad (2.96)$$

or in the integral form,

$$H(t) = H(0) + \int_0^t \left(\mu(u) - \frac{\sigma^2(u)}{2}\right)du + \int_0^t \sigma(u)dW(u). \qquad (2.97)$$

Coming back to H definition, we obtain the solution of the original SDE,

$$S(t) = S(0)\exp\left(\int_0^t \left(\mu(u) - \frac{\sigma^2(u)}{2}\right)du + \int_0^t \sigma(u)dW(u)\right). \qquad (2.98)$$

Equation (2.98) is of fundamental importance as it provides the exact value of $S(t)$ without the need to discretize the path using the SDE given by Eq. (2.95). This represents a huge gain in terms of computational time, and that is why usually Quants look for SDE with known solution when they need to model financial variables. For example, assuming time-independent diffusion and drift terms, $\mu(t) = \mu$ and $\sigma(t) = \sigma$, Eq. (2.98) becomes

$$S(t) = S(0)\exp\left(\left(\mu - \frac{\sigma^2}{2}\right)t + \sigma W(t)\right). \qquad (2.99)$$

So, in order to numerically simulate $S(t)$, one needs to generate Gaussian random numbers with zero mean and variance equal to t, in order to simulate $W(t)$, and then plug it into Eq. (2.99). On the contrary, if the solution of the SDE was unknown, one would need to discretize in a proper way the time interval $[0, t]$, and exploit a discretized version of the SDE. This implies that, in order to obtain a single realization $S(t)$, one should perform a number of operations equal to the number of discretized time intervals.

As a final remark about this example, we observe that, in order to estimate the expected value of $S(t)$, it would be tempting to consider only the drift part of Eq. (2.99) just dropping the Brownian noise. Unfortunately, this approach neglects the role of the Itô's second-order derivatives, and it conduces to the wrong result. The correct one is given by

$$\mathbb{E}(S(T)) = S(0)e^{\mu T}. \qquad (2.100)$$

To conclude this section, we provide the generalization of the Itô's lemma in two dimensions. In this case, the cross derivatives with respect to the random variables must also be taken into consideration:

$$df(t, X, Y) = \frac{\partial f(t, X, Y)}{\partial t} dt + \frac{\partial f(t, X, Y)}{\partial x} dX + \frac{\partial f(t, X, Y)}{\partial y} dY$$

$$+ \frac{1}{2} \frac{\partial^2 f(t, X, Y)}{\partial^2 x} d[X, X] + \frac{1}{2} \frac{\partial^2 f(t, X, Y)}{\partial^2 y} d[Y, Y]$$

$$+ \frac{\partial^2 f(t, X, Y)}{\partial x \partial y} d[X, Y], \tag{2.101}$$

where we did not indicate t and ω dependencies in the random variables X, Y in order to simplify notation.

2.15. Some Very Useful Theorems

In this section, we describe some very useful theorems that are used in many QF applications. Here, we report their statements in quite an informal way, and we focus on their intuitive meaning and their relevance in financial applications.

2.15.1. *Martingale Representation Theorem*

In Section 2.10, we observed that a very important feature of the Brownian motion is that each path $W(t, \omega)$ must be continuous in time. This would imply that each random variable $W(t + dt, \omega)$ at a given time $t + dt$ cannot lie too far from the previous one $W(t, \omega)$. Informally speaking, this tells us that the noise we introduce between t and $t + dt$ is not "strong" enough to completely destroy the $W(t, \omega)$-state as there is still a (continuous) connection with the state $W(t + dt, \omega)$. This observation provides us a good insight in terms of representation. In fact, even if we cannot obtain a better forecast of $W(t + dt, \omega)$ than the one provided by $W(t, \omega)$ (remember that the Brownian motion is a martingale), we could expect to *represent* $W(t + dt, \omega)$ by *another* random process, say $\hat{W}(t, \omega)$ multiplied by an appropriate function $\phi(t)$ which is known at time t.

This is exactly the statement of the *martingale representation theorem* [5]: given a probability space $(\Omega, \mathcal{F}, \mathbb{P})$, a filtration $\mathcal{F}(t)$ generated by[g]

---

[g]By *generated*, we mean that the information in $\mathcal{F}(t)$ is the one that can be obtained by the knowledge of $W(u)$ at time t.

the Brownian motion $\hat{W}(t)$ and another stochastic process $V(t)$, that is a martingale with respect to the same filtration, there is a process $\phi(t)$ known at time t such that

$$V(t) = V(0) + \int_0^t \phi(u) \mathrm{d}\hat{W}(u). \tag{2.102}$$

We observe that in our theorem statement we referred to a generic martingale stochastic process $V(t)$ that also includes a generic Gaussian process.

From a practical point of view, this theorem is of fundamental importance as it allows one to cancel the noise implicitly in a stochastic function without the need to properly forecast its future variations but just using another noise source. In a financial context, one could assume that $V(t)$ and $\hat{W}(t)$ are two assets prices that are both martingales under the same filtration. From martingale representation theorem, one can cancel the price variation $V(t) - V(0)$ continuously buying and selling the amount $-\phi(u)$ of the other asset represented by the process $\hat{W}(t)$. As implied by Eq. (2.102), the latter trading strategy would produce a total variation equal to $V(0) - V(t)$, completely offsetting the price variation of the asset $V(t)$.

This way to cancel price variations of an asset is called *hedging-strategy*, and it is at the base of rational pricing theory that will be further discussed in Chapter 4.

To conclude this section, we observe that the martingale representation theorem can easily be generalized considering a d-dimensional Brownian motion $\hat{W}^d(t,\omega) = (\hat{W}_1(t,\omega), \dots, \hat{W}_d(t,\omega))$ and a martingale $V(t)$ as above. Then, there exists a d-dimensional process $\phi^d(t) = (\phi_1(t), \dots, \phi_d(t))$ known at time t, such that

$$V(t) = V(0) + \int_0^t \phi(u) \cdot \mathrm{d}\hat{W}(u), \tag{2.103}$$

where the symbol $\cdot$ stands for the scalar product between the two arrays. As a simple example, we consider two stochastic processes:

$$S(t) = S_0 \exp\left(-\frac{\sigma_S^2}{2}t + \sigma_S W(t)\right),$$

$$V(t) = \sigma_V W(t), \tag{2.104}$$

where S_0, σ_S, σ_V are three positive real numbers. As $S(t)$ and $V(t)$ both depend on $W(t)$, they have the same filtration $\mathcal{F}(t)$. As a consequence, one can apply the martingale representation theorem in order to cancel $S(t)$

variations using $V(t)$ and choosing

$$\phi(t) = \frac{\sigma_S}{\sigma_V} S(t). \tag{2.105}$$

In fact, from equations of $S(t)$ and $V(t)$, we know that

$$S(t) - S(0) = \int_0^t \mathrm{d}S(u) = \int_0^t S(u)\sigma_S \mathrm{d}W(u),$$

$$\mathrm{d}W(t) = \frac{\mathrm{d}V(t)}{\sigma_V}. \tag{2.106}$$

So, substituting the second equation in the first one, we obtain Eq. (2.103), where $\phi(t)$ is defined by Eq. (2.105).

2.15.2. *Feynman–Kac Theorem*

In this chapter, we introduce some mathematical instruments that will be useful in order to model financial problems. By doing so, we implicitly choose a stochastic framework as the natural environment for our purposes. Actually, this is not the only possible choice, and one could wonder if there are others frameworks that could be more adequate to our aims. Obviously, it is very difficult to find a general answer to this question, and probably a suitable solution does not exist. On the other side, there exists a very interesting theorem, known as *Feynman–Kac theorem* [1] that establishes a clear link between the stochastic processes and the mathematics of the partial differential equations (PDEs). In fact, it can be proved that a solution of a PDE can be obtained considering an SDE and estimating the expected value. More formally, if $\mu(t,x), \sigma(t,x)$, and $\phi(t,x)$ are functions smooth enough, the solution of the PDE,

$$\frac{\partial V(t,x)}{\partial t} + \mu(t,x)\frac{\partial V(t,x)}{\partial x} + \frac{1}{2}\sigma^2(t,x)\frac{\partial^2 V(t,x)}{\partial x^2} = rV(t,x), \tag{2.107}$$

with the terminal condition,

$$V(T,x) = \phi(x), \tag{2.108}$$

can be obtained by estimating the expectation,

$$V(t,x) = \exp[-r(T-t)]\mathbb{E}(\phi(X(T,\omega))|\mathcal{F}_t). \tag{2.109}$$

Here, $X(t, \omega)$ is a stochastic process with the following dynamics:

$$dX(t, \omega) = \mu(t, X(t, \omega))dt + \sigma(t, X(t, \omega))dW(t, \omega), \qquad (2.110)$$

where $\mathcal{F}_t$ is the filtration generated by the Brownian motion $W(t, \omega)$, and the measure under which the expectation is estimated is the one implied by the SDE.

From this theorem, it is evident that the function $\mu(t, x)$ of the PDE represents the drift part of the SDE, while the function $\sigma(t, x)$ is the diffusive part. From the mathematical point of view, we observe that in the PDE, $\mu(t, x)$ and $\sigma(t, x)$ are deterministic functions of the dummy variable x, while in the SDE, x is substituted by the stochastic random variable $X(t, \omega)$.

In Chapter 4, we will show that Eq. (2.109) represents the solution of the risk-neutral pricing problem and it can be obtained by approaching the problem with both PDE and SDE.

Feynman–Kac theorem represents a very interesting example of how stochastic (SDE) and deterministic (PDE) approaches can interact and can be exploited in order to find solutions to practical problems.

2.15.3. *Radon–Nikodym Theorem*

In Section 2.2, we observed that the probability of an event is not an absolute concept, but it depends on the way we measure it, or in other words, on the probability measure definition. As a consequence, from a theoretical point of view, the same event could have different probabilities (and different expectations), depending on the measure we chose to measure it. At this point, one could wonder if there is a way to switch from a probability measure to another one, and if it is possible to define some classes of equivalence of the probability measures. As probabilities are, after all, measures of "volumes", one could try to compare them, considering their ratio with respect to the events in the probability space. For example, considering two probability measures $\mathbb{P}_1$ and $\mathbb{P}_2$, one can take the ratio $\mathbb{P}_2(A)/\mathbb{P}_1(A), \forall A \in \mathcal{F}$, where $\mathcal{F}$ is the σ-algebra of the probability space. This ratio compares how we measure the same set A by two measures. The considered ratio is well-defined only if the denominator is not 0. As in probability theory, there could be events of zero probability, the only way to avoid divergences is to require that $\mathbb{P}_1(A) = 0$ *implies* that $\mathbb{P}_2(A) = 0$. If this is the case, we say that $\mathbb{P}_2(A)$ is *absolutely continuous* with respect to $\mathbb{P}_1(A)$. If the opposite is also true, the two probability measures agree on zero-probability events, and for this reason, they are said to be *equivalent*.

Once the pathological cases are avoided, we can state a theorem that allows one to convert a probability measure into another equivalent one: considering a probability space $(\Omega, \mathcal{F}, \mathbb{P})$ with filtration $\mathcal{F}_t$ and two equivalent measures $\mathbb{P}_1(\omega), \mathbb{P}_2(\omega)$, there exists a martingale $J(t, \omega)$ such that

$$\mathbb{P}_2(E) = \int_E J(t, \omega) \mathrm{d}\mathbb{P}_1(\omega) \qquad \forall E \in \mathcal{F}_t. \tag{2.111}$$

In other words, this theorem assures that it is always possible to switch from a measure to another one, considering an appropriate stochastic process $J(t, \omega)$ that can be proved to be a martingale. Writing Eq. (2.111) in differential terms, one can obtain a derivative-like notation for $J(t, \omega)$,

$$J(t, \omega) = \left. \frac{\mathrm{d}\mathbb{P}_2}{\mathrm{d}\mathbb{P}_1} \right|_{\mathcal{F}_t}. \tag{2.112}$$

For this similarity with derivative notation, $J(t, \omega)$ is usually called *Radon–Nikodym derivative* [1], and the above-mentioned theorem is called the *Radon–Nikodym theorem*. Radon–Nikodym derivative can be efficiently used in order to simplify the expectation estimation. In fact, one can obtain the expectation change-of-measure formula, simply starting from the expectation definition,

$$\mathbb{E}^{\mathbb{P}_2}(x) = \int_\Omega X(\omega) \mathrm{d}\mathbb{P}_2(\omega)$$

$$= \int_\Omega X(\omega) \frac{\mathrm{d}\mathbb{P}_2}{\mathrm{d}\mathbb{P}_1} \mathrm{d}\mathbb{P}_1(\omega)$$

$$= \mathbb{E}^{\mathbb{P}_1}\left(x \frac{\mathrm{d}\mathbb{P}_2}{\mathrm{d}\mathbb{P}_1} \right). \tag{2.113}$$

The last equation expresses the original expectation under the measure $\mathbb{P}_2$ as the expectation of a new random variable $x\frac{\mathrm{d}\mathbb{P}_2}{\mathrm{d}\mathbb{P}_1}$ under the probability $\mathbb{P}_1$. As Eq. (2.113) holds for *any* measure $\mathbb{P}_1$ (equivalent to $\mathbb{P}_2$), one could exploit this formula, and choose a suitable equivalent measure in order to simplify the expectation calculation.

Equation (2.113) can be further extended to the conditional expectation case. In order to do this, one needs to consider Eq. (2.17) that defines the conditional expectation, and apply the change of measure,

$$\int_A \mathbb{E}^{\mathbb{P}_1}\left(X\frac{d\mathbb{P}_2}{d\mathbb{P}_1}\Big|\mathcal{F}_t\right)(\omega)d\mathbb{P}_1(\omega) = \int_A X(\omega)\frac{d\mathbb{P}_2}{d\mathbb{P}_1}d\mathbb{P}_1(\omega)$$

$$= \int_A X(\omega)d\mathbb{P}_2(\omega)$$

$$= \int_A \mathbb{E}^{\mathbb{P}_2}\left(X|\mathcal{F}_t\right)(\omega)d\mathbb{P}_2(\omega), \qquad (2.114)$$

where all the equations hold $\forall A \in \mathcal{F}_t$. Rewriting Eq. (2.114) in differential terms, we obtain

$$\mathbb{E}^{\mathbb{P}_1}\left(X\frac{d\mathbb{P}_2}{d\mathbb{P}_1}\Big|\mathcal{F}_t\right)(\omega)d\mathbb{P}_1\Big|_{\mathcal{F}_t}(\omega) = \mathbb{E}^{\mathbb{P}_2}\left(X|\mathcal{F}_t\right)(\omega)d\mathbb{P}_2\Big|_{\mathcal{F}_t}(\omega), \qquad (2.115)$$

where by the notation $\Big|_{\mathcal{F}_t}$, we mean that the equation holds in any subset of the filtration $\mathcal{F}_t$. Rearranging the terms, we get

$$\mathbb{E}^{\mathbb{P}_2}\left(X|\mathcal{F}_t\right) = \frac{\mathbb{E}^{\mathbb{P}_1}\left(X\frac{d\mathbb{P}_2}{d\mathbb{P}_1}\Big|\mathcal{F}_t\right)}{\frac{d\mathbb{P}_2}{d\mathbb{P}_1}\Big|_{\mathcal{F}_t}}, \qquad (2.116)$$

our *conditional expectation change-of-measure formula*. As a last passage, we observe that the denominator on the right-hand side of Eq. (2.116) can be interpreted as a conditional expectation of the Radon–Nikodym derivatives. After all, $J(t,\omega)$ in Eq. (2.112) was defined as a random process, so the expectation is well defined. Using again Eq. (2.17), one obtains

$$\int_A \mathbb{E}^{\mathbb{P}_1}\left(\frac{d\mathbb{P}_2}{d\mathbb{P}_1}\Big|\mathcal{F}_t\right)d\mathbb{P}_1(\omega) = \int_A d\mathbb{P}_2(\omega) \qquad (2.117)$$

or

$$\mathbb{E}^{\mathbb{P}_1}\left(\frac{d\mathbb{P}_2}{d\mathbb{P}_1}\Big|\mathcal{F}_t\right) = \frac{d\mathbb{P}_2}{d\mathbb{P}_1}\Big|_{\mathcal{F}_t}. \qquad (2.118)$$

2.15.4. *Girsanov's Theorem*

Once the change-of-measure is defined, one could wonder how the Radon–Nikodym derivative looks like. The answer to this question is provided by *Girsanov's theorem* [7] that describes how to move from a measure to another one, just acting on the drift term of an SDE. We consider two smooth enough functions $\mu(X(t,\omega))$ and $\sigma(X(t,\omega))$, representing respectively the drift and the diffusion coefficient of the SDE,

$$dX(t,\omega) = \mu(X(t,\omega))dt + \sigma(X(t,\omega))dW(t,\omega), \qquad (2.119)$$

under a given measure $\mathbb{P}_1(\omega)$. We want to find a new measure $\mathbb{P}_2$, under which the stochastic process follows the same SDE as in Eq. (2.119), but with a different drift $\mu^*(t,\omega)$,

$$dX(t,\omega) = \mu^*(X(t,\omega))dt + \sigma(X(t,\omega))dW^*(t,\omega), \qquad (2.120)$$

where $W^*(t,\omega)$ is a Brownian motion under the measure $\mathbb{P}_2$. Assuming that the ratio $(\mu^*(x) - \mu(x))/\sigma(x)$ is bounded, Girsanov theorem provides the analytic expression of the Radon–Nikodym derivative for the required change of measure

$$\left. \frac{d\mathbb{P}_2}{d\mathbb{P}_1} \right|_{\mathcal{F}_t} (\omega) = \exp\left[-\frac{1}{2} \int_0^t \left(\frac{\mu(X(u,\omega)) - \mu^*(X(u,\omega))}{\sigma(X(u,\omega))} \right)^2 du \right.$$
$$\left. - \int_0^t \left(\frac{\mu(X(u,\omega)) - \mu^*(X(u,\omega))}{\sigma(X(u,\omega))} \right) dW(u,\omega) \right]. \qquad (2.121)$$

Girsanov theorem assures that the new measure $\mathbb{P}_2$ is equivalent to the old one $\mathbb{P}_1$, and that the new Brownian motion $W^*(t,\omega)$ is given by

$$dW^*(t,\omega) = \left(\frac{\mu(X(u,\omega)) - \mu^*(X(u,\omega))}{\sigma(X(u,\omega))} \right) dt + dW(t,\omega). \qquad (2.122)$$

Girsanov theorem is of fundamental importance in the option pricing theory because, as will be shown in Chapter 4, the price of an instrument is invariant in a wide family of *equivalent* probability measures. As a consequence, Girsanov theorem allows one to *choose* the most convenient measure just modifying the drift of the SDE, being sure that the results obtained from the model also hold under other equivalent measures. In Chapter 4, we will show that this property can be exploited to avoid arbitrage opportunities.

To conclude this section, we observe that it is possible to generalize Girsanov theorem to the multi-dimensional case as it was done in the case of the martingale representation theorem (Section 2.15.1). In this case, we consider the d-dimensional Brownian motion $W^d(t,\omega) = (W_1(t,\omega),\dots,W_d(t,\omega))$ with vector $\Theta^d(t) = (\Theta_1(t),\dots,\Theta_d(t))$ representing the d-dimensional version of the ratio $(\mu^*(x) - \mu(x))/\sigma(x)$ (with the exception of a minus sign) considered above. Requiring that the norm,

$$||\Theta(t)|| \equiv \left(\sum_{i=1}^{d} \Theta_i^2(t) \right)^{\frac{1}{2}}, \qquad (2.123)$$

be bounded, that is, in line with the previous requirement on the ratio $(\mu^*(x) - \mu(x))/\sigma(x)$, multi-dimensional Girsanov theorem assures that the Radon–Nikodym derivative for the change of measure is given by

$$\frac{d\mathbb{P}_2}{d\mathbb{P}_1}\bigg|_{\mathcal{F}_t} (\omega) = \exp\left[-\frac{1}{2} \int_0^t ||\Theta(u)||^2 du \right.$$

$$\left. - \int_0^t \Theta(u) \cdot dW(u, \omega) \right]. \qquad (2.124)$$

Analogous to Eq. (2.122), it is possible to express the d-dimensional Brownian motion in the new measure as

$$dW^*(t, \omega) = \Theta(t)dt + dW(t, \omega). \qquad (2.125)$$

Chapter 3

The Pricing of Financial Derivatives — The Replica Approach

3.1. Introduction — The Pricing of Financial Derivatives

In this chapter, we deal with the problem of the rational pricing of financial contracts, and we show how the mathematical framework presented in the previous chapter can be exploited in order to obtain a consistent theory that aims to provide a rational approach to estimate the fair value of a contract and a hedging strategy for the risk mitigation. Even if this theory is self-consistent and quite fascinating, as it mixes financial concepts like absence of arbitrage and mathematical tools like martingale property of stochastic processes, it relies on strong hypotheses which should be considered more as an exception of standard market conditions than normality. If we reflect on this point, we should realize that this aspect is quite recurrent in mathematical modeling phenomena and, in our opinion, in general the hypotheses of a model should be regarded as a smart way to organize and to order concepts that otherwise are too complex to be understood. Ironically also from a more practical point of view, when we develop a model, we should pay more attention to the assumptions made than on the results obtained by the model itself, as the deeper is our understanding of our hypotheses, the lower is the risk of losing money in using such a model in practical situations.

A typical example can be obtained considering one of the most famous models still in use for the pricing of financial derivatives, i.e. the Black and Scholes model [8]. We will discuss all the features of this very simple model in Chapter 5, and it will be pointed out how the pricing of financial derivatives can be obtained with the estimation of only one free parameter, the so-called implied volatility. The hypotheses on which this model relies are quite strong and, in many cases, not justified as it can be shown from an empirical point of view. Nonetheless, the model is still used in many practical situations to price financial derivatives. In our opinion, this is the right choice as in this way it is possible to squeeze all the uncertainty related to the complexity of a financial system in a single and compressed number (the volatility) that can be easily understood and managed by traders avoiding the difficulties in modeling subtle features of financial systems that could be hardly estimated if not with a huge statistical error. Obviously, the price to pay for such a simple model is the need for managing the model hypotheses that in this case could lead to arbitrage opportunities and unwanted risks that eventually culminate in catastrophic events.

From this point of view, the best pricing model should be regarded as the one that can provide more insights about its weaknesses.

3.2.   What Kind of Model are We Looking For?

Every modeling activity should start answering this simple question. Typically, real-world situations are represented by awfully complex systems and, as was already pointed out in the introduction, in order to obtain manageable models, we are forced to introduce some simplification, neglecting minor features and dependencies that we assume do not affect considerably the final results of our model. In order to do this, it is necessary from the very beginning to know what are our final goals and the precision required to achieve them. From a pricing perspective, we can identify at least two main goals that should be achieved by a good model:

- To provide coherent pricing routines (of the whole portfolio) in order to avoid arbitrage opportunities between financial instruments that belong to the same portfolio (or, in general, available in the market).
- To provide a hedging strategy that minimizes the profit and loss (P&L) fluctuations and assures an expected P&L proportional to the risk-free rate.

In the following, we are going to discuss these points in more detail. In the classical option pricing framework, the two goals are achieved leveraging on the Martingale Representation Theorem already described in Section 2.15.1. In more complex (and realistic) frameworks, people sometimes focus only on one side of the problem, i.e. the pricing of financial instruments or the definition of optimal hedging strategies, developing different models to achieve both tasks independently. In this case, model coherence is usually analyzed by ex-post analyses.

3.2.1. *Coherence of the Pricing Methodology*

Let us assume that you are running a big financial institution and you have a huge portfolio composed by different assets. As a manager, you are interested in knowing what is the actual value of your portfolio at every time, as you want to understand if you are gaining or losing money and define new trading strategy on the basis of your recent P&L performances. In order to do this, you need to know the price of each instrument in your portfolio.

For many instruments, this is a very simple task and you do not need a model at all: it is enough to look at market quotes of your favorite broker of each financial instrument and you can immediately know what could be your realized P&L if you closed all the positions in your portfolio. Obviously, as it is just a theoretical calculation this does not mean that you actually realize that P&L until you do not decide to sell all the financial instruments; for this reason, often people talk about *unrealized* P&L. The strong hypothesis underlying this approach is that you will be able to close your position exactly at the price you look at the monitor of your personal computer; in doing so, you are neglecting, for example, that if the size of your order is very large, there cannot be enough counterparts in the market ready to satisfy your request (i.e. to buy your assets), so you could be forced to close your position at a different, less convenient price. Anyway, if the market is sufficiently liquid and the size of the order is not too large, the available market quotes are generally considered a good proxy for the P&L estimation. We stress that in this case, we are not making any assumption on the future evolution of market prices or on what should be the theoretical fair price of a financial instrument.

The situation changes a little if, for some reason, the market quotes were not readily available. For example, this situation is quite common when the trading is done over-the-counter (OTC), i.e. directly between two parties without any supervision of the exchange. In this case, the liquidity level is not assured to be high enough, so that the price, if available, cannot be considered a good proxy for P&L estimation. As a consequence, it is necessary to find a different way to price the financial instrument, and a model is required.

As suggested in Ref. [9], the simple approach to solve this kind of situation would be to look at similar reference instruments with known price and define an interpolation procedure to determine the price for the instrument that "lies between". The general hypothesis underlying this method is that instruments with similar characteristics should also have similar prices that can be represented by some smooth function. This interpolating smooth function should be determined by our mathematical model, and it represents the final goal of our analysis.

At this point, a very natural question is what is the meaning of smooth function from a financial point of view. By smooth function we mean that the price of the modeled instrument should not change too much with respect to the prices of the reference instruments, so that it would not be *possible* to implement a trading strategy on different portfolio instruments that could assure a positive expected P&L with a *zero probability* of losing money, i.e. it would not be possible to find an arbitrage opportunity. We will come back on this with a more formal definition of arbitrage opportunity in the following sections. Here we observe that our no-arbitrage condition is not very restrictive as its definition involves theoretical events with zero probability (or equivalently events with probability equal to one). As a consequence, there could be many possible definitions of *smooth functions* to interpolate similar prices in order to obtain the unknown price. As we will see in the following, a way to work around this arbitrariness is to consider a perfect *replica* of the instrument with other financial assets with known price; in this case it can be shown that, because of no-arbitrage condition, the (unknown) price of the financial instrument and the price of its replica have to be exactly the same.

In any case, whatever is the pricing method considered, our general requirement is that all prices obtained for our portfolio are coherent in the sense that it should not be possible to find out an arbitrage opportunity simply considering the prices generated by our models. In other words, we want to be sure that the appraisal of the unrealized P&L is the best estimation that can be obtained, and it is not possible to extract additional P&L from our portfolio with zero risk. In most of the cases, the development of a model (or many different models) that is able to price any financial instruments in a coherent way is not an easy task and, in general, it is possible to obtain coherent pricing method only for similar instruments that belong to the same asset class.

3.2.2. Hedging Strategies

As was already stated above, the coherence of the pricing methodology represents only one goal that should be achieved by a good pricing framework. Another typical requirement for a model is that it should be able to provide a hedging strategy that could cancel out, or at least, minimize the risk related to the financial instrument that we want to price. The reason for this requirement is twofold. From a theoretical point of view, the requirement of a hedging strategy is related to the replica approach described in the previous section: as the pricing method is based on the replica of the instrument of interest with other financial instruments, it should be possible to hedge the risk of the first instrument by its replica with the opposite sign. If we are not able to replicate the instrument by a hedging strategy, our theoretical framework would not be self-consistent.

From a more practical point of view, the knowledge of an effective hedging strategy implies the knowledge of the main risk drivers of your financial instrument. In other words, if you are able to replicate a given financial instrument with another (more fundamental) one, it means that you can represent all the risks related to the first instrument as a proper aggregation of the risks related to each financial instrument used for the replica. From this perspective, you can actually *explain* the P&L derived from that financial instrument as a sum of the P&Ls of its components (given by the replica).

In practical situations, hedging strategies are implemented at portfolio level as well as P&L explanation activities; in general, it is not possible to perfectly hedge the portfolio performance and in most of the cases an unexplained P&L is still present in the portfolio after hedging. A good pricing framework should be able to minimize this unexplained part.

3.3. Are We Able to do the Same? Replica Pricing

In this section, we describe a first standard approach to the rational pricing that is based on the replica of an instrument payoff, with other more fundamental instruments of known price.

As already outlined in the previous section, a very simple way to price something is to look at *similar* products of known price and try to extrapolate the fair price of the object you want to buy or sell. This is quite common also in your daily life, when you want to buy a new laptop, for example. You look at laptops with *similar* performances and you choose the one you think has the best ratio between price and technical features. Obviously, there is an intrinsic uncertainty about what we mean for *similar*, so different customers could choose different laptops because they estimate the similarity relations in a different way.[a]

In order to get rid of this uncertainty, one could look for *exactly* the same instrument, i.e. an instrument with the same characteristics of the original instrument: in this case, one could reasonably expect that the two prices should be the same. Why? Because otherwise one could buy the cheaper instrument and then sell it at the higher price, generating money from nothing, i.e. taking money from someone who sold the instrument at a cheaper price for free. The sentence *to generate money from nothing* implicitly means there are no risks to be taken into consideration when we put in place the buy and sell strategy previously described, otherwise, the word *nothing* should be replaced by *risk*. In particular, here we assume that the market is liquid enough so that we can buy or sell financial instruments contextually. As already stated in the previous chapter, the opportunity of realizing a profit without any risks is called *arbitrage* opportunity; from the previous example, the following definition should be clear:

One-Price Law: In an arbitrage-free market, two exactly equal instruments have the same price.

This very simple law gives us a very important hint about how to obtain a rational pricing routine. What we need is:

- To assume that there are no arbitrage opportunities in the market.
- To replicate a financial instrument with other (already priced) instruments. The unknown price is given by the cost of the replica.

Let us think a little bit on the last sentence. What we are assuming, through no arbitrage arguments, is that the (fair) price of a financial instrument is actually a cost; if everybody agrees with the price that is published in the market at a given time, everybody implicitly agrees on the cost that one has to face to replicate that instrument. This could be quite puzzling at first sight as if the price of an instrument in a liquid market is rather well determined and unique[b]; it could be expected that the costs of the

---

[a]So far, we are not assuming that customers act in a perfectly rational way, as irrationality can be understood as a not obvious way to build similarity relations.

[b]Here we are neglecting the so-called bid-ask spread and the dependence of the price on the volume of the transaction.

replica should strongly depend on the financial institution and its ability to manage fees; for example, its cost of funding or transaction expenses. In any case, one could think about these differences in commissions as additional transaction costs included as a mark-up on the price and assume that the price of an instrument is an overall average of the costs of the replica on which everybody agrees.

At this point, a very natural question could be: What do we exactly mean by *equal* instruments? As we assumed that the arbitrage opportunity is *never* possible, it is enough to require that the two instruments have the same payoff at maturity to be sure that they always have the same price. This consideration clarifies what we exactly mean by a *replica* strategy: it is an investment strategy involving different fundamental financial instruments with known price that is able to replicate the payoff of another instrument (with unknown price) at maturity.

In this approach, we are implicitly assuming that there is a relation between the fundamental financial instruments (that are assumed to be part of our economy) and the instrument of the unknown price, that for this reason is called *contingent claim*. If there exists a strategy that can replicate the payoff of the contingent claim, the latter is called *attainable*.

The last element required to complete our replica setup is a condition on the strategies that can be considered in the replica. A trivial strategy that can replicate any payoff could be to do whatever you want and at maturity add or take the money you need to exactly match the payoff of the instrument. This trivial strategy is undoubtedly able to replicate the payoff, but it requires that an amount of money (that typically depends on future market conditions) must be abruptly injected in the portfolio. In order to avoid all of this, in the following, we will require that the replica strategies have to be *self-financing*, i.e. no-money injection is allowed from outside after the starting time. Before concluding this introduction on replica pricing, we summarize the algorithm obtained for pricing a contingent claim:

- Assume that there are no arbitrage opportunities in the market.
- Find a self-financing strategy that can replicate the final payoff of the contingent claim with other, more fundamental, instruments.
- Estimate the cost of the replica: by the law of one price it is the price of your financial instrument.

In the following paragraphs, we will describe all these steps in a more formal way, and we will show some examples of application.

3.3.1. *Bank Account*

For every non-trivial trading strategy, it is a very natural to find a safe place to put the money when you need to decrease your positions in risky assets; symmetrically, the strategy could require to borrow money in order to take position in risky assets. In our theoretical framework, we assume that the activity of borrowing and lending money is done by the so-called *bank account* and that the dynamics of one unit of money in the bank account at $t = 0$ can be described by the following differential equation:

$$dS^0(t) = r(t)S^0(t)dt, \qquad (3.1)$$

where $S^0(t)$ represents the value of the money at time t and $r(t)$ is the rate at time t that we receive or pay, respectively, if we lend or borrow money from the account, assuming that no-risks are related to this kind of investment; for this reason, $r(t)$ is called *risk-free rate*. This bank account modeling is based on some underlying assumption that we make explicit in the following:

- *The bank account cannot default.* This assumption is not explicit in Eq. (3.1), but it is implicitly assumed if we define $r(t)$ as a risk-free rate. In particular, as we are assuming that no-risks are related to an investment in the bank account, from a financial perspective, one should require that $r(t)$ is less (or equal) than the minimum rate available in the market, as any risky investment would imply a risk premium that should be reflected in the rate.
- *The borrowing and lending rates are equal or, at least, their difference is negligible for our purposes.* This is quite a strong assumption as, in principle, the difference between the funding cost paid by a financial institution to borrow money and the rate it decides to lend its money could represent the main driver of its business. In any case, as already outlined in the previous section, the main goal of our model is to obtain a price on which all market participants can agree, so our model variables cannot depend on any specific business activity, and should represent an average cost which everybody accepts. Anyway, also from an "average" perspective, differences between borrowing and lending rate can be found in the market; in the following, we will assume that these differences are mainly related to risky assets, so they can be implicitly assumed to be zero as long as we define the bank account as a risk-free asset.

- *You can put and receive back your money from the account continuously in time.* This is a very fundamental assumption as it implies that there is no liquidity risk in this investment.
- *In an arbitrage-free world, all existing bank accounts are equivalent; if there are two risk-free investments their dynamics can be described by Eq. (3.1) with the same $r(t)$ function.* This is a simple application of the law of one price for a risk-free investment.
- *The risk-free rate depends on time.* In particular, we are not trying to model it as an explicit function of other macroeconomic variables.

On the other side, we observe that we did not make explicit hypotheses on the domain of $r(t)$ and, in particular, we did not assume that $r(t) > 0$. In the past years, this additional requirement was understood as natural constrain because of the financial intuition that nobody is inclined to pay in order to lend money; on the contrary, one may require the payment of an interest. In these years, empirical evidence shows that this constrain is no longer correct as the price of many financial contracts currently exchanged in the market implies a negative interest rate (see Section 7.1). Fortunately, the theory shown in these paragraphs does not require the positivity of the interest rate and it can be efficiently applied in the current market situation.

As a consequence of the dynamics postulated by Eq. (3.1), a unit of money invested in the bank account at $t_0 = 0$ worth at time t:

$$S^0(t) = \exp\left(\int_0^t r(s)\mathrm{d}s\right). \tag{3.2}$$

From a mathematical perspective, Eq. (3.2) tells us that we are introducing from the very beginning in our economy an exponential behavior that could lead to extreme scenarios as time goes to infinity. In our opinion, these kinds of divergences are hardly compatible with social systems and should be regarded as a warning about how our modeling assumptions can faithfully represent reality. On the other hand, $r(t)$ is typically very small, ranging from 0.01 to 0.1, considering 1 year as unit of time, so the exponential behavior can be efficiently approximated by the linear expansion:

$$\exp(r(t)t) \sim 1 + r(t)t. \tag{3.3}$$

As a consequence, the exponential behavior can be significantly different from the linear expansion only on very large time scales. On the contrary, considering a small time scale, Eq. (3.2) can be understood as a way to continuously accrue interests on a short time horizon. For example, considering a time discretization $\{t_1, \ldots, t_n\}$, $\Delta t = t_{i+1} - t_i$ and $r(t) = r$, we have:

$$S^0(t + \Delta t) \sim S^0(t)(1 + r\Delta t)$$

$$S^0(t + 2\Delta t) \sim S^0(t + \Delta t)(1 + r\Delta t) = S^0(t)(1 + r\Delta t)(1 + r\Delta t) \qquad (3.4)$$

$$\vdots$$

So, if we partition a finite time variation ΔT in n sections, and we let n go to infinity, we obtain:

$$S^0\left(t + n\frac{\Delta T}{n}\right) = S^0(t)\left(1 + r\frac{\Delta T}{n}\right)^n \rightarrow S^0(t)\exp(r\Delta T). \qquad (3.5)$$

From this simple example, we showed that the dynamics postulated in Eq. (3.1) represents a continuous reinvestment of the starting money, and all the gain is generated by the interests in the same account. This strategy would imply an exponentially increasing amount of the money in time; as already observed, given the small magnitude of the interest rate, this exponential behavior is not very different from a linear behavior on small time scale.

The bank account is a very fundamental element in our theoretical framework as it gives us information about how money changes its value in time. For example, assume that you are going to receive (for sure) a given amount of money, let us say $100,000\$$, in a year and you want to estimate what is the value of that money today. A naive answer would be that as you are going to receive that amount for sure, no risk is related to this investment, so the value of this investment today is the same, i.e. $100,000\$$. On the other hand, if you received $100,000\$$ today, you could put all the money in a bank account and after 1 year have $V(T) = 100,000\$ \exp\left(\int_0^T r(s)\mathrm{d}s\right)$ that is different from $100,000\$$ with the exception of the trivial case $r = 0$. In particular, if $r(t) > 0 \ \forall t$, $V(T) > 100,000\$$, it would always be better to receive all the money today than wait 1 year for it. In general, one can define the value of the money today $V(t)$ as the amount of money to be invested today in a bank account in order to *replicate* the future cash flow $V(T)$:

$$V(t)\exp\left(\int_s^T r(s)\mathrm{d}s\right) = V(T), \qquad (3.6)$$

or, in a more compact way

$$V(t) = D(t, T)V(T), \tag{3.7}$$

where

$$D(t, T) \equiv \frac{S^0(t)}{S^0(T)} = \exp\left(-\int_t^T r(s)\mathrm{d}s\right) \tag{3.8}$$

is the so-called *discount factor*, and allows to propagate backward the price of an instrument from future time T to current time t.

In order to make clear that today's value of a future cash flow cannot be equal to the value of the cash flow in the future, previously, we assumed that $r(t) > 0 \; \forall t$; this is a quite reasonable hypothesis as in general one expects that lending money implies the requirement of a *positive* interest. Actually, from a historical series analysis, it can be shown that in some cases (as nowadays), interest rates can be negative. In any case, this situation is not incompatible with our theoretical framework, even if quite counterintuitive. In fact, as we assumed that the market is arbitrage-free and all the risk-free investments are equivalent, all investments not paying (or receiving) the risk-free rate cannot be considered risk-free. From this point of view, even if you do not invest your money in any asset or bank account, you are making a *risky* investment, so you could lose or make money. In particular, you will lose money (with respect to a risk-free investment) if the hypothesis of positive risk-free rate holds, on the contrary you will make money.

The assumption that the bank account is a risk-free asset is questionable in real market situations; from a practical point of view, people look for the best *approximation* of a risk-free investment and consider it as a bank account (see for example [10, 11] for collateral agreement modeling). From this perspective, the bank account represents the *best low-risk alternative* that one can find in the market: all the asset prices will be valued as an investment with respect to this low-risk alternative; the bank account is considered as a reference asset, or in technical terms a *numeraire*.

In some cases, the cash flows in and out of the bank account compensate themselves, and there is no need to look for a bank account proxy. An example of this situation is given in Ref. [12], where the existence of a bank account is *postulated* but it does not affect pricing results. On the contrary, it is shown that the more natural account to consider as a reference asset is the so-called *collateral* account. Similar results are obtained in Ref. [10], where the collateral account is assumed to be a proxy of the bank account.

3.3.2. Self-Financing Strategies

In Section 3.3, we discussed how to price a financial instrument by means of a very simple approach: the replica.

To put it in mathematical terms, first we define our financial world as a probability space $(\Omega, \mathcal{F}, \mathcal{F}_t, \mathbb{P})$ where there are $n + 1$ financial instruments (including the bank account) that we will consider as fundamental and non-dividend paying. Their price can be modeled by a $n + 1$ dimensional stochastic process $S = \{S(t)|0 \leq t \leq T\}$ where $S^0(t), S^1(t), \ldots, S^n(t)$ are its positive components and T is our time horizon, so that $0 \leq t \leq T$. In addition, we require that $S(t)$ should be represented by an adapted process added to a martingale process.

Starting from this framework, we define a *trading strategy* as a $(n + 1)$-dimensional process $\phi = \{\phi(t)|0 \leq t \leq T\}$ whose components $\phi^0(t), \phi^1(t), \ldots, \phi^n(t)$ are locally bounded and adapted. The *value process* associated with the trading strategy ϕ is defined by:

$$V(t, \phi) \equiv \phi(t)S(t) = \sum_{k=0}^{n} \phi^k(t)S^k(t), \quad 0 \leq t \leq T \tag{3.9}$$

and the *gain process* is given by:

$$G(t; \phi) \equiv \int_0^t \phi(u)\mathrm{d}S(t) = \sum_{k=0}^{n} \int_0^t \phi^k(u)\mathrm{d}S^k(u), \quad 0 \leq t \leq T. \tag{3.10}$$

In this framework, we can define a large set of trading strategies involving many financial instruments to obtain the replica. As our mathematical framework is based on continuous stochastic processes, the gain estimation is obtained by an integration over the time, considering the price variations in differential terms. In addition, we observe that we required the ϕ components to be bounded in order to control the size of the investment in each fundamental assets, avoiding divergences, and predictable as at any time we need to know exactly the amount of each fundamental instrument in our portfolio. In this framework, $V(t; \phi)$ represents the market value of the portfolio, implied by the strategy ϕ at time t; i.e. the amount of money we would receive at time t if we closed all positions. This interpretation is correct as long as we assumed that the fundamental assets $S(t)$ are liquid, so we can close the position at every time t, and their price is known and it is given by t. For the same reasons, $G(t; \phi)$ represents the gain related to the strategy ϕ up to time t.

We define a trading strategy ϕ *self-financing* if $V(\phi) \geq 0$ and

$$V(t; \phi) = V(0; \phi) + G(t; \phi). \qquad (3.11)$$

As mentioned above, the intuitive interpretation of Eq. (3.11) is that the value of the strategy at time t is given by the money we invested at the very beginning $t = 0$ plus all the profits and losses generated by the strategy itself; we are not allowed to inject (or take) additional money in (from) our portfolio in order to change its value. With respect to self-financing strategies, a very important theorem (see Refs. [13, 14]) tells us that it is possible to introduce a discount factor in Eq. (3.11) without any change in the meaning of the equation; in other words, given a trading strategy ϕ, ϕ is self-financing *if and only if*

$$D(0, t)V(t; \phi) = V(0; \phi) + \sum_{k=0}^{n} \int_{0}^{t} \phi^k(u) \mathrm{d}\left(D(0, u)S^k(u)\right). \qquad (3.12)$$

We remark that the *if and only if* expression means that Eqs. (3.11) and (3.12) are equivalent, so one can take Eq. (3.12) as a definition for self-financing strategies.

Equation (3.12) tells us that we can move backward in time and discount all the cash flows of a trading strategy, but this *time-travel* does not affect our definition of self-financing strategy itself. Intuitively, the reason why the self-financing relation still holds after the introduction of the discount factor is that the exponential nature of the discount factor does not perturb significantly the original equation (technically speaking, the discount factor is a bounded variation function) and so it acts like a normalizing factor that changes the *scale*, i.e. the amount, of the quantities involved in the equation but not the qualitative properties. This normalizing nature of the discount factor is very helpful in our mathematical framework because it can be used to *control* the divergences in the equations of the fundamental assets in order to obtain a martingale behavior. In fact, as for positive risk-free rates, the discount factor is a decreasing exponential function of time, exponentially diverging dynamics for the fundamentals assets can be normalized and transformed in martingale just multiplying them by discount factor. This aspect will be discussed in the risk-neutral pricing framework in Section 4.1.

3.3.3. *Arbitrage*

The absence of arbitrage is one of the most fundamental hypotheses in the rational pricing framework and it is, in our opinion, one of the most fascinating and intellectually intriguing concepts of the financial world. From this hypothesis, different constraints on the price of an instrument can be derived and exploited to obtain a mathematical equation providing, once solved, the price of financial instruments. From a practical point of view, the absence of arbitrage should be regarded as a theoretical assumption that holds in the long run, as arbitrage opportunities, when existing, are immediately exploited by market participants and thus disappear in a small amount of time.

As already mentioned above, arbitrage is defined as the possibility to implement a (self-financing) trading strategy on different portfolio instruments ensuring a positive expected P&L with zero probability of losing money. In formulas, if $V(t; \phi)$ is the value of self-financing strategy, ϕ generates an arbitrage if

$$V(0; \phi) = 0,$$

$$\mathbb{P}(V(T; \phi) \geq 0) = 1, \qquad (3.13)$$

$$\mathbb{P}(V(T; \phi) > 0) > 0,$$

where $\mathbb{P}$ is the *real-world* probability measure related to $V(t; \phi)$. The first requirement condition tells us the initial cost of the strategy is equal to zero. This is not a very restrictive condition as by *cost of the strategy*, we mean the money we have to take from our pocket in order to put it in place. As in the previous section, we assumed that there exists a bank account that can lend or borrow you money, you can always convert a strategy in a zero-cost starting strategy using the bank account. The second requirement in Eq. (3.13) tells us that the final value (at time T) of our strategy is never negative, so your final P&L can be positive or equal to zero. The last requirement tells us that there is a non-zero probability to realize a profit using this strategy. In an informal way, these three conditions are telling us that we have the opportunity to realize a gain without any drawbacks, i.e. it is not possible to lose money.

A very peculiar aspect of arbitrage definition is that only zero and one probabilities are involved; or in other words, is not required a deep knowledge about the properties of the chosen measure. It is only required that the measure is able to characterize certain and impossible events. From a modeling point of view, this is a huge advantage as we are not forced to choose the real-world probability, but we are allowed to select *any* measure that is equivalent to the real-world one. This is a very important aspect as it is very difficult to obtain for financial systems a good model for the future real-world probability distribution and the reason is quite simple: if we were able to forecast the future real-world distribution of prices, we could become rich! As a consequence of the arbitrage definition, we can choose any equivalent probability measure and we can be sure that if no-arbitrage is not possible in the new measure, it is also not possible in the real-world one.

Thus, to ensure to have no-arbitrage opportunities in our system, we need a statement that falsifies at least one of the three conditions in Eq. (3.13). In this case, the trick is to require that the expectation of the P&L of our strategy at time T is equal to 0. This is tantamount to the requirement that

$$\mathbb{E}\left[V(T;\phi)\right] = 0, \tag{3.14}$$

given that $V(0;\phi) = 0$. This equation falsifies the second requirement of Eq. (3.13), as in order to obtain the expected value of $V(T;\phi)$ equal to 0, it is necessary that some values of $V(t;\phi)$ are negative. In other words, the requirement that the expected P&L is equal to zero introduces some risk in our trading strategy that cannot be considered as an arbitrage. What is very interesting from our point of view is that the expectation of the P&L can be taken in an *arbitrary*, and more convenient, equivalent measure.

From a financial perspective, the no-arbitrage requirement $\forall t$ assures a kind of coherence of the prices through time as two instruments that have the same value (the same payoff) in the future T cannot have a different price at any time $t < T$, otherwise it would be possible to make an arbitrage, as already explained in Section 3.3. In fact, one could buy the cheaper instrument at price $I_c(t)$ and sell the more expensive one at price $I_e(t)$ at time t and close the position at time T. As at time T, the two instruments have the same payoff and the net P&L is given by

$$\text{P\&L} = (I_c(T) - I_c(t)) - (I_e(T) - I_e(t)) = I_e(t) - I_c(t) > 0. \tag{3.15}$$

With this simple example, we showed that in an arbitrage-free market, the price must be unique for instruments with the same final payoff. This fact is the base of the replica approach to price financial instruments. In the following, we will give some practical examples on how to use this approach to price simple financial instruments.

3.3.4. *Replica Examples*

In the preceding sections, we have already discussed the importance of the three main concepts in Quantitative Finance (QF): arbitrage, self-financing portfolios and the pricing by the replica approach. In this section, we will give some examples to exploit these concepts, and in particular, the replica approach to price different financial instruments. Motivated by this goal, we will also give a short description of the main financial instruments which will also be used in the following sections as fundamental blocks for examples of practical applications.

In a nutshell, the replica approach is based on the following steps (Fig. 3.1):

- Choose a target instrument whose price is unknown and write its payoff at maturity.
- Define a self-financing strategy based on fundamental instruments whose price is known so that the sum of the fundamental instruments payoffs is equal to the target payoff at maturity. This is the replica strategy.
- The price of the strategy at different times is known by definition as the prices of the instruments are known.
- The price of the target instrument at any time is equal to the price of the replica due to the no-arbitrage condition, i.e. law of one price.

The power of this approach i.e. in the fact that we did not make any modeling assumption on the dynamics of the fundamental instruments; for this

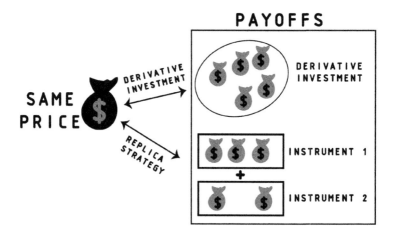

Fig. 3.1.   A representation of the replica algorithm to price an instrument.

reason, this approach is called model-free. The difficult part is *how* to define the self-financing replica strategy.

3.3.4.1. *Zero Coupon Bond*

As a first example of the replica approach, we consider one of the most fundamental instruments of the interest-rate market: the zero coupon bond (ZCB), as suggested by the name. This contract guarantees its holder the payment of one unit of currency at maturity T without any intermediate payments, hence the name of the contract. In financial terms, it means that by buying this contract, you are lending money to the issuer of the bond that will pay you back at maturity. Obviously, as you are lending money, you are entitled to ask for an interest. In this case, as the final payoff is fixed and equal to one unit of currency, in order to include a positive interest into the contract, the amount of money lent at time $t < T$ should be less than one and it should represent the price of the bond. From a financial point of view, one could expect that the lower the probability of receiving back the money at maturity, the higher the interest rate required as a risk premium should be, as a consequence, the price of the bond should be lower. From this point of view, the price of the bond should be related to the *default probability* of the issuer; for this reason, this kind of bond is called *defaultable bond*.

In our example, we want to exclude this default component in the pricing, so we will consider only bonds issued by default-free entities, or more realistically, bonds where the default event is negligible. In this case, one could wrongly think that the price of the bond should be equal to the amount of money to be received at maturity, i.e. equal to one, as no-risks affect our investment. Actually, we already discussed this case in Section 3.3.1 where we showed that the value of the money is not constant in time, as one can receive a risk-free interest just investing money into a bank account, as a consequence, the price of the bond must be lower than one. In order to simplify the problem, we assume that the risk-free rate is constant in time, i.e. $r(t) = r$. As a consequence, our bank account becomes deterministic and its dynamics is given by

$$S^0(t) = \exp(rt). \qquad (3.16)$$

In particular, $S^0(t)$ represents the amount of money we have in our account at time t if we invested a unit of currency into the bank account at $t = 0$. Now, we choose the bank account as our fundamental instrument and we look for a replica of our payoff. In this case, the amount of money ϕ we have to invest into the bank account at time $t = 0$ in order to match the ZCB payoff at time T is given by the equation,

$$\phi \exp(rT) = 1, \tag{3.17}$$

so that

$$\phi = \exp(-rT). \tag{3.18}$$

By this simple equation, we obtained our replica of the ZCB payoff:

- At $t = 0$, invest a $\phi = \exp(-rT)$ amount of money into the bank account.
- Wait until maturity T, where the money in the bank account will be

$$M(T) = \phi S^0(T) = \exp(-rT)\exp(rT) = 1 \tag{3.19}$$

that is exactly the payoff of the ZCB.

Given that we obtained the replica and the price of the bank account is known at time t, the price of the ZCB is

$$P(t, T) = \phi S^0(t) = \exp(-rT)\exp(rt) = \exp(-r(T - t)). \tag{3.20}$$

This example shows how the replica approach can be used to price financial instruments in a very simple way starting from very fundamental hypotheses. In our case, we observe that quite a hidden hypothesis is related to the time at which we can receive our money back. In fact, if the ZCB pays back all the money only at maturity, the same is not true for the bank account as, by definition, it is possible to receive back our money at any time $t \leq T$. So, from this point of view, our replica is equivalent to the ZCB investment as long as we neglect the opportunity to interrupt the investment *before* maturity or, in other words, the liquidity risk. In the last years, as a consequence of the financial crisis, liquidity played a fundamental role, and the market participants also started to include this aspect in their pricing models. In the last chapter of this book, we will give some hint on how to deal with this topic.

In addition, as already explained above, the price of a bond should also reflect its default probability that was completely neglected in the previous argument. In this respect, Eq. (3.20) should be regarded as a practical formula to transform a unit of currency of a future cash flow in a constant continuously compounded interest rate r multiplied by the right time increment $T - t$. In this way, it is possible to compare in a simple way different investments that generate cash flows over different time horizons. Following the same lines, ZCB should be regarded as a theoretical instrument acting as a reference in the interest-rate market.

3.3.4.2. *Depo Rates*

Another hypothesis we made in setting up our replica strategy for the ZCB was that we assumed a constant risk-free rate. This hypothesis is quite reasonable for all the financial products that are not strongly linked to interest-rate world, i.e. their payoff is not strongly affected by a small variation in an interest rate, like in the case of equity products. Obviously, this is not true for interest rates linked products as the interest-rate modeling assumptions could strongly affect the final price of our instrument. For example, if we assume that the risk-free rate is not constant but stochastic, the replica of the ZCB cannot be done by a single investment in the bank account as its value at maturity is a stochastic variable and it cannot be assured that it will be equal to one, as required by ZCB payoff. In this case, the replica of the ZCB payoff can be achieved considering a *deposit*. This financial instrument, investing an initial capital N at time t, pays back the capital plus an interest when at maturity T as given by the following expression:

$$Depo(t,T) = N(1 + (T - t)R(t,T)), \qquad (3.21)$$

where $R(t,T)$ is the interest rate fixed at time t for the maturity T. Assuming that the deposit contract is liquid enough, the replica argument shown in Section 3.3.4.1 can also be applied here simply replacing the bank account with the deposit contract. In this case, the price of the ZCB is given by

$$P(t,T) = \frac{1}{1 + (T - t)R(t,T)}. \qquad (3.22)$$

Equation (3.22) is a ZCB pricing formula that exploits the replica argument and generates a relation between the bond and the deposit contract. As already stated above, Eq. (3.22) is equivalent to Eq. (3.20) if we assume that the risk-free rate is constant and we neglect in the pricing the default probability related to the payer of the interest in the deposit contract. In this case, we obtained a formula that allows to discount future cash flows by a factor that depends on a *constant linearly compounded* interest rate $R(t,T)$ multiplied by the right time increment $T - t$.

3.3.4.3. *Forward*

As an additional example of how to price a financial instrument using the replica approach, we consider the *forward*. The forward is a financial contract where the two parties agree to respectively buy and sell a given quantity N of an *underlying* instrument at a given price K, called *strike*, at a given time T in the future, called *maturity* of the contract. In general, this kind of instruments is called *derivative* (or *contingent claim*) as the price of the instrument is strongly related to the price of an underlying instrument; so it should be possible to *derive* its price starting from the price of the underlying.

In our case, the forward contract is the derivative instrument and quite obviously its price will depend on the price at maturity of the exchanged underlying, $S(T)$. For example, if $S(T)$ is larger than the agreed strike price K, the buyer of the forward can take advantage from this situation as one can buy the underlying at price K and sell it in the market at price $S(T)$ realizing a positive P&L. On the contrary, if $S(T) < K$, the buyer of the forward realizes a loss as it is *forced*, by the contract, to buy the underlying at the agreed price K that is larger than the market quotation. In summary, the payoff of the forward $Fwd_K(S,T)$ at maturity T can be expressed as (Fig. 3.2)

$$Fwd_K(S,T) = N(S(T) - K). \tag{3.23}$$

In the following, for all these kinds of payoffs, we will neglect the subscript K whenever the strike is clear by the context. Equation (3.23) gives us our target payoff that should be reproduced by our replica strategy. Firstly, we observe that N is only a scaling factor that simply magnifies or reduces the effects of the difference between $S(T)$ and K on the right-hand side of the equation; without any loss of generality, we can safely assume that $N = 1$. Now, we can decompose our replica strategy in two parts: the first

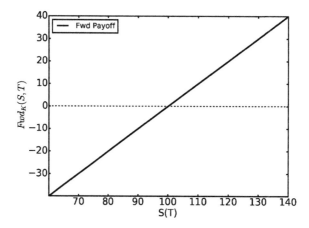

Fig. 3.2.   We show the forward payoff as a function of the price of the underlying at maturity $S(T)$ for a fixed strike $K = 100$ and $N = 1$.

one aims to replicate a constant value $-K$ at maturity, the second aims to obtain $S(T)$ at maturity. In the previous paragraph, we have already observed that the easiest way to receive a given amount of money K in the future is to buy a ZCB with notional equal to K. In our case, as we want $-K$, i.e. we want to *pay* money, we need to sell a bond with maturity T and notional K. The second part of our strategy is quite trivial: in order to obtain $S(T)$ at maturity, we just need to buy the underlying at price $S(t)$ and deliver it at maturity. To sum up, our replica strategy is made of two fundamental instruments: a bond and the underlying instrument of the forward contract; by selling the former with notional equal to K and buying the latter, we are able to reproduce the same payoff of Eq. (3.23), i.e. the forward payoff. By the usual no-arbitrage reasons, the value of the forward at time $t < T$ is given by

$$Fwd_K(S(t), t) = N((S(t) - K \exp(-r(T - t)))), \qquad (3.24)$$

where we reintroduced the scaling factor N and we used Eq. (3.20) for the bond price at time t, assuming that the risk-free rate r is constant. Typically, the strike value K of the forward is set at inception so that the value of the contract is zero. By imposing this condition to Eq. (3.24), we obtain that the value of the forward at inception is given by

$$K = S(t = 0) \exp(rT). \qquad (3.25)$$

Also in this case, we could obtain the price of a derivative instrument just starting from no-arbitrage assumption and replica arguments. What is quite surprising at first sight is that we did not need to make *any* assumption on the future value of the underlying or its dynamics; we simply exploited the replica strategy machinery.

As done for the ZCB, we remark that some approximations have been made. For example, we assumed that the risk-free rate is constant so that we can equivalently invest money in the bank account or in a bond. In general, even if the risk-free-rate is not constant, our replica strategy still holds, provided that we substitute in our formula the price of the bank account with the bond price:

$$K = \frac{S(0)}{P(0, T)}. \qquad (3.26)$$

3.3.4.4. *Forward Rate Agreement*

As another application of the replica pricing, we consider a financial instrument in the interest-rate world, the so-called Forward Rate

Agreement (FRA). In an FRA, the two parties in the contract exchange an interest-rate payment for a given time period. In particular, at maturity T, one of the two parties pays to the other one a fixed rate K and receives a floating payment based on an interest rate $L(S, T)$ that was fixed at time $S < T$ and it refers to the period $[S, T]$. In this way, at time $t < S < T$, it is possible to lock-in a future and *unknown* interest rate L, exchanging it with a *known* interest rate K. The length of the time period for which the interest rate is paid is given by the difference between T and S in an appropriate day-count convention. To summarize, the FRA is a contract that involves three time instants:

- the current time t,
- the fixing time $S > t$, that is the time when the floating interest rate is fixed,
- the maturity time $T > S > t$, that is the time when the payments between the two parties is made.

Formally, at time T, one receives $N(T - S)K$ units of currency and pays $N(T - S)L(S, T)$, where N is the contract nominal value and acts as a scaling factor of the investment. So, the payoff at $t = T$ is

$$FRA(t = T; S, T, N, K, \tau) = N(T - S)(K - L(S, T)). \qquad (3.27)$$

Even if in principle $L(S, T)$ could represent *any* interest rate, in practical situations, $L(S, T)$ refers to an Interbank Offered Rate (IBOR), i.e. the interest rate at which banks lend to and borrow from one another in the interbank market. In particular, the London Interbank Offered Rate (LIBOR) is one of the most representative interbank rates and it is estimated by taking the average interest rate (for interbank lending and borrowing money) among the leading banks in London. As a consequence, in order to put in place our replica strategy, we firstly need to choose an interest-rate product that could safely represent the LIBOR. In a classical QF framework, one typically assumes that this representation can be achieved considering interbank deposits:

$$P(t, T) = \frac{1}{1 + L(t, T)(T - S)} \qquad (3.28)$$

that can be obtained by identifying in Eq. (3.22) the deposit rate $R(t, T)$ with the LIBOR $L(t, T)$. Substituting the last expression in

Eq. (3.27), we get:

$$FRA(t = T; S, T, N, K, \tau) = N\left((T - S)K + 1 - \frac{1}{P(S,T)}\right). \quad (3.29)$$

Within the last expression, we decomposed FRA payoff in two parts:

- A deterministic cash flow given by $N((T-S)K+1)$ that is fixed when the contact is written and known without uncertainty at any time $t < S < T$. As we already observed in the preceding sections, a future known cash flow can be replicated by a ZCB investment with the same maturity of the FRA. For the usual replica arguments, the price of this cash flow at time t is given by $N((T - S)K + 1)P(t, S)$.
- A cash flow given by $-\frac{1}{P(S,T)}$ that can be known without uncertainty only for $t \geq S$, i.e. after the fixing date. In order to reproduce this cash flow at $t < S$, one can sell a bond with maturity S, so that at time $t = S$, one has to pay a unit of currency. In order to make this payment, one could think of selling another bond with maturity T and notional equal to $1/P(S,T)$. In this way, as the price of the bond at time S is exactly $P(S,T)$, one receives at time S a unit of currency. At maturity $t = T$, as a consequence of this investment, one has to pay $1/P(S,T)$ unit of currency that perfectly replicates the second part of the payoff.

Using the replica strategy, we obtain the price of the FRA at time $t < S$. In particular, we look for the K value that makes the contract worthless at time t:

$$FRA(t; S, T, N, K, \tau) = NP(t, S)(1 + K(T - S)) - P(t, T) = 0. \quad (3.30)$$

Solving Eq. (3.30), one obtains the so-called *par*-rate of the FRA:

$$K = \frac{1}{T - S}\left(\frac{P(t, T)}{P(t, S)} - 1\right). \quad (3.31)$$

Also in this case, using only replica arguments, we could obtain a pricing formula that prices a very fundamental instrument of the interest-rate market. Here, the replica requires two operations in the market at different times, t and S, in order to hedge LIBOR fluctuations and obtain a self-financing strategy. Remarkably enough, we were able to obtain a value of K that implies no-exchange of money at inception as the value of the contract is zero; this makes the FRA quite appealing in the market. This demonstration for the pricing of the FRA belongs to a classical QF framework and it was commonly accepted before recent post-2007 financial crisis. After this, it was proved that some of the hypotheses underlying this approach are not valid anymore and the whole framework must be reorganized in order to take into consideration new sources of risk that were traditionally neglected. We will discuss this new framework in Chapter 7.

Despite these adjustments, one could expect that generalizing the above-described approach and making use of the replica strategy at many different times, it should be possible to price more complex financial instruments; this is essentially the aim of the risk-neutral pricing approach that will be described in Section 4.1.

3.4. Derivatives: Call, Put and Digital Options

In this section, we give a brief description of the so-called option derivatives with special focus on the most fundamental ones: call and put options. These kinds of contracts have nonlinear payoffs and for this reason, it is hard to replicate them through standard linear instruments. As a consequence, they can be used as a starting point to introduce a general approach that will be described in Section 4.1 allowing the pricing of complex financial instruments. In general, options are contracts that give the holder the *possibility* to do something in the future with respect to the underlying contract, only if market conditions are favorable; on the contrary, if market conditions are not favorable, the holder of the option could decide not to exercise the right. From this simple definition, it should be clear that in option contracts, there is a sort of asymmetry between the holder and the seller of the option. In particular, the latter has a passive role as the nature of the payoff is essentially defined by the decisions of the former. This asymmetry of the contract allows the seller the option to ask for the payment of a *premium* that represents essentially the value of the option itself. From a mathematical perspective, the asymmetry introduces a nonlinear effect that is reflected in the mathematical framework and increases its complexity.

Option derivatives can be organized in classes considering the futures times at which exercise is possible:

- European options: the option can be exercised only at a pre-determined future time (maturity).
- American options: the option can be exercised at *any* time up to maturity.
- Bermudan options: the option can be exercised at a given set of future times; in other words, they are between European and American options, exactly as Bermuda is in between Europe and America.

The most fundamental option types are the so-called *call* and *put* options. In the following, we will focus on the *European* calls and puts, but most of the consideration also apply for other kinds of calls and puts. A European call (put) option is a derivative contract that gives the holder the opportunity (but not the obligation) to buy (sell) a given amount N of the underlying asset at maturity time T at a given strike price K determined when the contract is issued. At maturity T, the holder of the call (put) option can decide to buy (sell) the N amount of underlying at price K as a function of the market conditions.

In particular, for the holder of a call option, it is worthwhile to exercise one's right if the price of the underlying at maturity is larger than the strike price K. In this case, one could buy the underlying at price K and sell it at the market price $S(T) > K$ and realize a net gain equal to $S(T) - K$ that is the value of the call at maturity. On the contrary, if the market price of the underlying is less than K, there is no reason for the call holder to buy the underlying at price K as one could buy it at a cheaper price $S(T) < K$. In this case, the holder will not use the right of buying the underlying at price K, so the value of the call would be equal to zero. From a practical perspective, call and put contracts are quite useful in order to hedge the risks related to market movements over or below a given threshold with a relatively cheap cost.

We can summarize the payoff of the call option in the following short notation:

$$C_K(S,T) = \max(S(T) - K, 0), \tag{3.32}$$

where $C_K(S,T)$ is the price of the European call option with strike K at maturity. By a similar argument, the payoff of a European put option at maturity $\mathrm{Put}_K(T)$ is given by

$$\mathrm{Put}_K(S,T) = \max(K - S(T), 0). \tag{3.33}$$

In Fig. 3.3, we show the payoffs of the two instruments for a fixed strike $K = 100$. Equations (3.32) and (3.33) have a very fundamental role in our theoretical and mathematical framework as they summarize in a very short notation the deep nature of option contracts, i.e. the nonlinearity of their payoffs. In other words, our mathematical representation allows us to make a strong connection between the optionality of the contract and the nonlinearity of the problem. This fact should be understood as a warning message about the possibility of putting in place a replica strategy for these nonlinear payoffs. In fact, as a replica strategy is essentially a linear

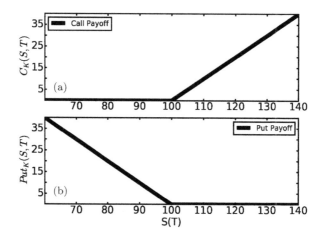

Fig. 3.3. We show the call (a) and put (b) payoffs as a function of the price of the underlying at maturity $S(T)$ for a fixed strike $K = 100$.

combination of financial instruments (mainly with linear payoffs), it should be clear that the replica of a nonlinear payoff is not an easy task. Actually, it will be shown in Section 4.1 that the replica of this payoff is still possible considering a dynamic and continuous adjustment of the positions of the underlying assets together with the requirement of a continuous dynamics of the underlying.

From a completely different perspective, call and put options could be understood as the building blocks for the replica and hedging of other, more complex, nonlinear payoffs; from this point of view, they *complete* the market for the hedging of nonlinear effects.

Often, call and put options are categorized by the difference between their strike price and the current spot price of the underlying:

- *at-the-money* options: the current spot price is equal to the strike price,
- *in-the-money* options: the spot price is higher than the strike price,
- *out-of-the-money* options: the spot price is below the strike price.

This classification is commonly accepted and will help us in identifying the most relevant situations for the option pricing.

Another example of European options is given by the so-called *digital option*. Its payoff splits the possible future scenarios of the underlying into two classes: scenarios above a given threshold K that represents the strike price and scenarios below it. As a consequence, the payoff only provides two outcomes: pay all the amount or nothing. A typical example is represented

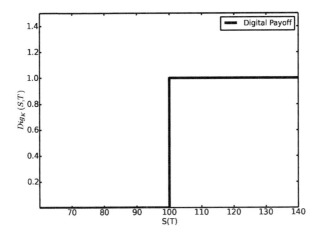

Fig. 3.4. We show the cash or nothing call payoff as a function of the price of the underlying at maturity $S(T)$ for a fixed strike $K = 100$.

by the *cash or nothing call* where payoff is given by

$$\text{Dig}_K(S,T) = \begin{cases} 1 & \text{if } S(T) > K, \\ 0 & \text{if } S(T) \leq K. \end{cases} \qquad (3.34)$$

As evident from the formula, in this case, you receive 1 if the underlying at maturity is larger than K and 0 otherwise. This payoff is represented in Fig. 3.4.

Also in this case, the price of the instrument can be obtained either considering the risk-neutral approach or realizing an approximation of its payoff combining nonlinear instruments like calls and puts. We will discuss both cases in Section 5.5.

Chapter 4

Risk-Neutral Pricing

4.1. Introduction

The replica strategy is a very powerful tool that allows one to price different financial instruments starting from very simple hypotheses, as no-arbitrage condition, and without any assumption on the dynamics of the underlyings. In particular, if the replica is feasible in the market, i.e. it involves only liquid and tradable assets, a misalignment between the market and the replica price would represent a gain opportunity without any risk (i.e. an arbitrage). As in the market, nobody wants to pay lunches for free,[a] the market price of a financial instrument is somehow forced to be linked to no-arbitrage price; if the no-arbitrage condition does not hold, someone could put in place an arbitrage strategy until the market price goes back to the no-arbitrage price. As a consequence, even if arbitrages are present in the market, this recursive mechanism assures that the price of a financial instrument in a liquid market cannot lie far from the no-arbitrage price for a long time. This behavior makes the replica approach very robust as it is essentially based on no-arbitrage condition.[b]

Unfortunately, the determination of the replica strategy is not always an easy task as in most of the cases, it would require a continuous trading

---

[a]This expression is quite common in Quantitative Finance (QF) theory and it can be regarded as an informal restatement of the no-arbitrage requirement.

[b]From a very practical point of view, we observe that an arbitrage can be actually exploited only if the mismatch between the arbitrage price and the no-arbitrage one is larger than the so-called bid-ask spread.

activity that cannot be determined *a priori* without assumptions on the dynamics of the underlying.

In general, one could look for a strategy able to *locally* replicate the profit and loss of the financial instrument of interest and continuously *adapt* it, reacting to market changes. This continuous adjustment to market movements represents a generalization of the replica approach discussed in Chapter 3, where only "static" replicas were described.

Intuitively, one could expect that, *if no abrupt market changes occur*, this adaptive strategy will be able to *hedge* all the risks related to market movements and thus replicate the final payoff. We stress from the very beginning that two conditions are required in order to proceed with this approach:

- The market movements of the instrument we want to use for the replica have to be locally continuous as our replica strategy can react to fluctuations with a little delay or, in other words, we do not expect any forecast of future market movements from our strategy. As a consequence, we have no chance to cancel out the market fluctuations if they are not continuously connected. From a mathematical perspective, this implies that Wiener processes are the most natural choice to model the dynamics of the underlying financial instrument; whenever their dynamics cannot properly represent the actual ones, some adjustment to the general theory would be required.

- The trading activity for the replica of the product should be continuous in time because of our local perspective. This hypothesis is quite unrealistic as transaction costs are, in general, increasing functions of the number of operations, so a continuous trading activity would imply a divergence of costs that would be reflected into an excessively high price of the instrument (remember that the price is the cost of the replica).

These two hypotheses allow one to generalize the replica approach in order to obtain a coherent mathematical framework where the replica pricing problem is transformed to the problem of estimating an expectation under a proper probability measure. In fact, it will be shown that the pricing formula of *any* contingent claim can be obtained estimating the expected value of its payoff under a probability measure called *risk-neutral*. Furthermore, the replica strategy of the payoff can be achieved through the estimation of a mathematical (partial) derivative.

We will refer to this approach as risk-neutral pricing from the name of the probability measure considered (Fig. 4.1).

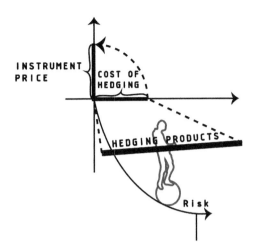

Fig. 4.1. A representation of the risk-neutral approach as a way to hedge the risks of an instrument by buying and selling other hedging products. The cost of the hedging must be equal to the price of the instrument.

4.1.1. *Setting the Stage*

Risk-neutral pricing derivation requires the analysis of different conceptual steps; in order to avoid getting lost during the derivation, we summarize what we already know about pricing and the mathematical elements that will be relevant to accomplish our task. We define a *derivation program* that will be used to keep in mind the big picture of the derivation.

From the preceding paragraphs, the key elements that could be useful are as follows:

- The no-arbitrage is defined considering only self-financing strategies which include a discount factor.
- The no-arbitrage definition is quite general and holds for a *class* of equivalent probabilities; we can choose for our derivation the most *convenient* distribution inside the class implied by the physical probability (i.e. the probability of the market) and results will also hold for it.
- To price something, we need to replicate its payoff with a self-financing strategy.

From a mathematical perspective, we have:

- The martingale representation theorem (see Section 2.15.1) assures the existence of a replica for martingale process with other martingale processes.

- Girsanov theorem tells us how to switch between two equivalent probability measures and how to obtain the desired drift.
- The exponential nature of the discount factor can be useful to cancel the drift component of the stochastic process and transform it in a martingale. In addition in many cases (especially for equity asset class), its randomness does not create any bind with other processes, so it can be included in the equations without taking into consideration the covariance structure (Eq. (3.12)).

In order to exploit all these key elements, we will consider in our derivation the following steps that will be clarified in the following sections:

- We model the underlying dynamics with a general stochastic differential equation. We have in mind to apply the martingale representation theorem, thus all sources of noise in financial markets have to be represented by a Wiener process.
- We guess a (pricing) function that relates the contingent claim payoff with its present value that is a martingale under a wide class of measures. This assures the existence of a replica strategy with the martingale representation theorem. We keep some degree of freedom in the definition of the measure that will be used in the next steps.
- In order to set up a replica strategy based on the underlying of our contingent claim, we *choose* a measure in the previous measure class that assures our replica strategy to be a martingale too, using Girsanov's theorem.
- Equating the expressions of the dynamics of the contingent claim and the replica portfolio, we can identify our dynamic replica strategy, i.e. how to dynamically replicate our contingent claim with the considered portfolio.
- As we can set up a replica strategy, the guessed function in the chosen measure is actually the correct one to use for the contingent claim pricing; therefore, we will be able to formulate a theoretical algorithm that can be applied whenever we need to price a contingent claim.

In the following, we describe each step of this recipe in detail in order to derive the risk-neutral pricing formula, discussing the main hypotheses introduced and their relevance for practical applications. We stress that, as our approach strongly relies on the martingale representation theorem, the theory should be adjusted in a suitable way when one wishes to introduce non-continuous processes for the dynamics description of the underlying.

4.1.2. *Point One: Modeling the Underlying*

In this section, we make some hypotheses on the (tradable) underlying dynamics. In order to assure the applicability of the martingale representation theorem, we consider a stochastic differential equation with Wiener increments; in particular, we assume that the dynamics of our underlying is given by

$$dS(t) = \mu(t)S(t)dt + \sigma(t)S(t)dW(t), \qquad (4.1)$$

where $\mu(t)$ and $\sigma(t)$ are two generic time-dependent functions representing, respectively, the drift of the underlying and its *volatility*, i.e. the diffusion part. From a financial perspective, we are assuming that our underlying has a dynamics similar to a bank account (see Eq. (3.1)) for what concerns the deterministic part (given by the dt term), but with a drift component $\mu(t)$ that in general can be different from the risk-free rate. Furthermore, we added a stochastic part depending on the Wiener process increment and multiplied by a volatility factor to represent the erratic dynamics typically followed by a financial instrument like an equity.

It is worthwhile to make some additional comments on what we stated so far:

- The price increment is proportional to the current value $S(t)$; thus, the lower the price, the lower the increment. In particular, it is assured that the price cannot be lower than zero, which is considered as a positive feature of the model as the asset value cannot be negative. From another point of view, when the asset value is zero, it implies that it is defaulted; as a consequence, in our representation, we are implicitly neglecting the default risk. In addition, in some cases, the positivity of the underlying is not suitable to obtain a good representation of reality. For example, at the time of writing, we observe that some interest rates implied by the market are negative; therefore, the model expressed by Eq. (4.1) should be adjusted, e.g. by proper shifts to take into account this empirical behavior.
- As we used a Wiener increment to model the noisy part of the financial market, we are implicitly assuming that the price time series is continuous in time, for $\mu(t), \sigma(t)$ smooth enough. Unfortunately, this is not in agreement with the empirical analysis of the time series: in most of the cases, the price dynamics is discontinuous with jumps and spikes. This behavior is usually modeled including into the equation an additional stochastic process typically based on a different distribution that generates discontinuous paths. As a consequence, in this case, the application

of the martingale representation theorem would not be straightforward and some additional hypotheses are required. In order to keep things simple, we will not discuss these aspects in this book, only considering underlyings where the jump component is negligible, like equity indexes.

- By the application of the Itô's lemma, it is possible to obtain the following expression for $z(t) \equiv \log(S(t))$:

$$dz(t) = \left(\mu(t) - \frac{\sigma(t)^2}{2}\right)dt + \sigma(t)dW(t). \tag{4.2}$$

This transformation allows us to better understand the meaning of the parameters in our equation. First of all, it should be observed that if the variation of the underlying over a small time increment is small (this is assured by our continuity assumption), the following approximation holds:

$$dz(t) = \log(S(t + dt)) - \log(S(t)) \sim \frac{S(t + dt) - S(t)}{S(t)}, \tag{4.3}$$

or, in other words, the log-variation of the underlying can be interpreted as the relative price increment. This suggests a very practical meaning for the $\sigma(t)$ parameter: it represents the standard deviation of the relative variation of prices.

- We considered a time-dependent drift term $\mu(t)$ that is in general different from the risk-free $r(t)$. In particular, from a financial point of view, one could think that the following relation holds: $\mu(t) > r(t)$. The rationale of this requirement is that $S(t)$ represents a risky investment because of the presence of the Wiener term, in contrast with the bank account that is, by definition, risk-free. As a consequence, a rational investor should expect a higher return for the risky investment, i.e. a higher drift term. This condition is hard to be verified from an empirical point of view, as, typically, the underlyings are very erratic. This generates a large instability in the statistical estimation of $\mu(t)$.

Given these considerations, Eq. (4.1) provides quite a general dynamics for the underlying that will be exploited in the following sections to find a replica of our contingent claim.

4.1.3. *Point Two: Guess a Pricing Formula*

In this section, we need to make an educated guess of a pricing formula for a contingent claim with the only requirement that this formula generates a martingale under quite a large class of probability measures. As this requirement is quite weak, we try to use financial intuition in order to make our guess more reliable. First, we observe that if we know for sure that we are going to receive a future cash flow $V(T)$ at time T, its present value $V(t)$ is simply a discounted cash flow:

$$V(t) = D(t, T)V(T), \tag{4.4}$$

where $D(t, T)$ is the discount factor between t and T. This relation can be reformulated using a more useful notation, exploiting discount factor definition (Eq. (3.8)):

$$D(0, t)V(t) = D(0, T)V(T). \tag{4.5}$$

If the future cash flow is not deterministic, we can assume that $V(T)$ is an $\mathcal{F}(T)$-measurable random variable, so that we have many possible future cash flows that should be taken into consideration, each weighted by its probability. So, quite a natural guess for a pricing function would be to take the discounted expected value of these future cash flows:

$$D(0, t)V(t) = \mathbb{E}[D(0, T)V(T)|\mathcal{F}(t)], \tag{4.6}$$

where we intentionally omitted the measure related to our expectation. Using iterated conditioning (Eq. (2.23)), it can be proved that Eq. (4.6) is a martingale with respect to the measure associated to the expectation:

$$\mathbb{E}[D(0, t)V(t)|\mathcal{F}(s)] = \mathbb{E}[\mathbb{E}[D(0, T)V(T)|\mathcal{F}(t)]|\mathcal{F}(s)]$$
$$= \mathbb{E}[D(0, T)V(T)|\mathcal{F}(s)]$$
$$= D(0, s)V(s). \tag{4.7}$$

As a consequence, Eq. (4.6) is a very good candidate for our risk-neutral pricing formula, as it is quite general (i.e. the probability measure can be arbitrarily fixed) and it assures that the discounted price is a martingale. If we assume that the filtration $\mathcal{F}(t)$ is generated by a Brownian motion $W(t)$, the martingale representation theorem holds and the random variable $V(t)$ can be represented as

$$D(0, t)V(t) = V(0) + \int_0^t \Gamma(u)dW(u) \qquad 0 \le t \le T. \tag{4.8}$$

In other words, through this theorem, it is ensured that the discounted value of our contingent claim at time t can be replicated by an initial position $V(0)$ and a function $\Gamma(t)$ that multiplies our Wiener process increments. In this context, the good news is that we have not yet specified the measure we need to use in order to price our contingent claim. Unfortunately, the bad news is that in general, we are not able to directly buy or sell in the market the Wiener process increment, so we need to make some hypotheses to define the relation between the Wiener increment and the underlying instrument we want to use to put in place our replica. This point will be discussed in the next section.

4.1.4. Point Three: Finding an Equivalent Martingale Measure

In order to find a replica strategy, we look for a relation between the underlying $S(t)$ and the value $V(t)$ of the contingent claim we want to price, exploiting Eq. (4.8). In particular, we consider the following portfolio:

- We start with a given amount of money X_0.
- We buy a $\Delta(t)$ amount of shares $S(t)$ of the underlying, where $\Delta(t)$ is a function of time that drives our replica strategy. When $\Delta(t)$ is negative, it means that we have to sell that shares amount.
- We invest the remaining part of our portfolio $X(t) - \Delta(t)S(t)$ into the bank account receiving a risk-free interest $r(t)$ over the infinitesimal time interval dt. Equivalently, if $X(t) - \Delta(t)S(t)$ is negative, we assume that we borrow from the bank account that amount of money at the same interest rate.

In general for our portfolio, we have the following dynamics:

$$dX(t) = \Delta(t)dS(t) + r(t)(X(t) - \Delta(t)S(t))dt, \qquad (4.9)$$

where the right-hand side represents the variation respectively due to fluctuations of the asset price and the bank account.

If we substitute Eq. (4.1) in Eq. (4.9), we obtain:

$$\begin{aligned}
dX(t) &= \Delta(t)dS(t) + r(t)(X(t) - \Delta(t)S(t))dt \\
&= \Delta(t)\left(\mu(t)S(t)dt + \sigma(t)S(t)dW(t)\right) \\
&\quad + r(t)\left(X(t) - \Delta(t)S(t)\right)dt \\
&= r(t)X(t)dt + \Delta(t)\left(\mu(t) - r(t)\right)S(t)dt + \Delta(t)\sigma(t)S(t)dW(t) \\
&= r(t)X(t)dt + \Delta(t)\sigma(t)S(t)\left(\theta(t)dt + dW(t)\right) \\
&= r(t)X(t)dt + \Delta(t)\sigma(t)S(t)dW^{\mathbb{Q}}(t), \qquad (4.10)
\end{aligned}$$

where we defined $\theta(t) \equiv (\mu(t) - r(t))/\sigma(t)$ and we applied Girsanov's theorem (Section 2.15.4) to the last term in the right-hand side of the equation, obtaining the new Wiener increment $\mathrm{d}W^{\mathbb{Q}}$ defined under a suitable measure $\mathbb{Q}$ that absorbs the $\theta(t)$ term. It is worthwhile to remark that the Girsanov theorem application actually consists in a change of the drift term of the underlying dynamics; in fact in the new measure, the dynamics of the underlying becomes

$$\mathrm{d}S(t) = r(t)S(t)\mathrm{d}t + \sigma(t)S(t)\mathrm{d}W^{\mathbb{Q}}(t). \tag{4.11}$$

We observe that the change of measure "amplitude" $\theta(t)$ depends on the difference between the actual drift of the underlying $\mu(t)$ and the risk-free rate $r(t)$ normalized by the instantaneous volatility of the market $\sigma(t)$, so it represents the excess of return expected from a risky investment, normalized by the risk of the investment itself and represented by the instantaneous volatility. For this reason, $\theta(t)$ is called *market price of risk*.

Furthermore, the $\mathbb{Q}$-measure assures that the portfolio dynamics is a martingale under it, as can be shown applying Itô's lemma to the discounted portfolio:

$$\mathrm{d}(D(0,t)X(t)) = D(0,t)r(t)X(t)\mathrm{d}t + D(0,t)\Delta(t)\sigma(t)S(t)\mathrm{d}W^{\mathbb{Q}}$$
$$-X(t)r(t)D(0,t)\mathrm{d}t,$$
$$\mathrm{d}(D(0,t)X(t) = D(0,t)\Delta(t)\sigma(t)S(t)\mathrm{d}W^{\mathbb{Q}}, \tag{4.12}$$

or, in integral terms,

$$D(0,t)X(t) = X(0) + \int_0^t \Delta(u)\sigma(u)D(0,u)S(u)\mathrm{d}W^{\mathbb{Q}}(u). \tag{4.13}$$

Following the same lines, it can be shown that the discounted underlying is a martingale too under the same measure:

$$\mathrm{d}(D(0,t)S(t)) = D(0,t)\mathrm{d}S(t) + S(t)\mathrm{d}D(0,t)$$
$$= D(0,t)\left(r(t)S(t)\mathrm{d}t + \sigma(t)S(t)\mathrm{d}W^{\mathbb{Q}}(t)\right)$$
$$- r(t)S(t)D(0,t)\mathrm{d}t$$
$$= \sigma(t)D(0,t)S(t)\mathrm{d}W^{\mathbb{Q}}(t). \tag{4.14}$$

Since the measure $\mathbb{Q}$ transforms the general portfolio dynamics in a driftless dynamics, it is called *risk-neutral* measure. The choice of this measure is of fundamental importance in our setup as it assures that the portfolio gain given by our strategy under this particular measure is zero.[c] This assures that no positive gains are possible if you do not want to *risk* your money, i.e. the probability of losing money cannot be zero. In this respect, we have already observed that the zero expectation condition implies the no-arbitrage hypothesis; furthermore, if this condition is fulfilled, it holds in *every* equivalent measure we choose. In other words, the choice of the risk-neutral measure assures that it is not possible to find an arbitrage in our strategy in *every* equivalent measure and in particular in the physical market measure. As a consequence, if we found a strategy in the risk-neutral measure able to replicate a given payoff (applying martingale representation), we would be sure that this strategy is arbitrage-free and thus gives the price of the contingent claim at any time.

At the moment, we only showed that the replica strategy exists and it is arbitrage-free. In the next section, we are going to find the actual strategy to be put in place to obtain the replica.

4.1.5. *Point Four: Defining the Replica Strategy*

In order to find the replica strategy, all we need to do is to compare Eq. (4.8) with Eq. (4.13). If we assume that at the starting point $X(0) = V(0)$, then the two equations are equal, i.e. our strategy replicates the contingent claim payoff if

$$\Delta(t) = \frac{\Gamma(t)}{\sigma(t)D(0,t)S(t)} \tag{4.15}$$

that tells us that there exists a solution to our problem, but it is still not clear as the $\Gamma(x)$ function is in general not given. In order to find out a concrete expression for $\Delta(t)$, we need to apply Itô's lemma to the discounted expression of our pricing Eq. (4.6):

$$d(D(0,t)V(t)) = D(0,t)dV(t) + V(t)dD(0,t)$$
$$= D(0,t)\left(\frac{\partial V}{\partial t}dt + \frac{\partial V}{\partial S}dS + \frac{1}{2}\sigma^2(S,t)\frac{\partial^2 V}{\partial S^2}dt\right)$$
$$- r(t)V(t)D(0,t)dt$$

---

[c]This is evident if we take the expectation of Eq. (4.12) and we interpret the left-hand side of the equation as an infinitesimal gain.

$$= \left(D(0,t)\frac{\partial V}{\partial t} + D(0,t)\frac{\partial V}{\partial S}\mu(t)S(t) \right.$$

$$\left. - r(t)V(t)D(0,t) + \frac{1}{2}\sigma^2(S,t)\frac{\partial^2 V}{\partial S^2} \right) dt$$

$$+ D(0,t)\frac{\partial V}{\partial S}\sigma(t)S(t)dW(t)$$

$$= D(0,t)\frac{\partial V}{\partial S}\sigma(t)S(t)dW(t), \tag{4.16}$$

where we considered for $S(t)$ the usual dynamics (Eq. (4.1)) and we made use of the martingale property of Eq. (4.6) in the last expression. Comparing Eqs. (4.12) and (4.16), we obtain

$$\Delta(t) = \frac{\partial V}{\partial S} \tag{4.17}$$

that gives us concrete information about how to dynamically hedge our product. In fact, assuming that we are able to estimate the price of a contingent claim through Eq. (4.6) in the risk-neutral measure already defined in Section 4.1.4, the replica of the payoff can be obtained by taking the derivative of the price with respect to the underlying and buying that amount of shares in the market. The remaining money has to be invested in the bank account at the current risk-free rate. In this way, we are reproducing the strategy expressed by Eq. (4.9) that we know is a perfect replica of the contingent claim. As time goes by, we will be required to adjust the amount of shares that we need to own in our portfolio in order to replicate the value of our contingent claim. From this point of view, it is quite intuitive that the amount of underlying required for the replica will depend on how much the value of the contingent claim changes as a consequence of the variation of the underlying that is exactly the concept of mathematical derivative. As a consequence of this interpretation, the $\Delta(t)$ function is sometimes called *sensitivity* to the underlying. At this point, it is worthwhile to stress that, within this theoretical framework, we are able to obtain the price and the replica of a contingent claim without the need to make any forecast on the future behavior of the underlying product, but just by exploiting all the information available at time t to dynamically hedge the product. It turns out that all this information is actually included in the sensitivity of the product.

4.1.6. *Point Five: The Pricing Algorithm*

Given all the previous points, we are now ready to formulate a general algorithm that can be applied to price a contingent claim, given the assumptions previously described about the continuity of the underlying dynamics (Fig. 4.2). The algorithm can be summarized in the following steps:

- Define a stochastic differential equation for the underlying and define an appropriate measure $\mathbb{Q}$ (risk-neutral measure) such that the discounted dynamics is a martingale as in Eq. (4.14). This is tantamount to substituting the drift term of the equation of the underlying with the risk-free rate.

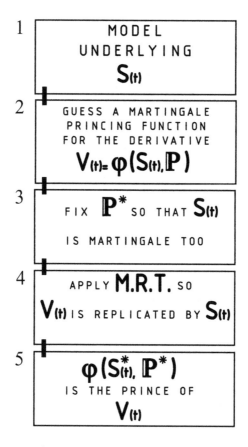

Fig. 4.2.   A representation of the risk-neutral algorithm.

- Express the final payoff of the contingent claim as a function $\phi(S(T), T)$ at the maturity T.
- Solve Eq. (4.6) under the previously defined measure to evaluate the price of the contingent claim:

$$D(0,t)V(t) = \mathbb{E}^{\mathbb{Q}}[D(0,T)V(T)|\mathcal{F}(t)].\qquad(4.18)$$

- Estimate the derivative of the price $V(t)$ with respect to the underlying in order to obtain the amount of underlying to buy (or sell) for the replica.

This algorithm is very powerful as it only requires the estimation of an expectation in order to price a wide range of financial instruments. In particular, one can couple this algorithm with Monte carlo simulations where underlying scenarios are numerically simulated by a computer and the sample mean is used to approximate the expected value in Eq. (4.18). In this setup, one needs to generate scenarios of $S(T)$ only once and then estimate the sample mean of the payoffs of all the financial instruments related to that underlying. As in general, the scenario generation is the most time-consuming part, this setup remarkably improves the computation performance.

On the other side, using Feynman–Kac theorem (Section 2.15.2), one can transform Eq. (4.18) in a partial differential equation (PDE), than solve it using standard numerical and analytic approaches to PDE.

Finally, we observe that one can price new products by just adding the new payoff into the pricing environment disregarding the modeling issue of the underlying dynamics, whenever the discounted dynamics is a martingale under risk-neutral probability. This is the main aspect that differentiates the pricing modeling world with respect to the risk world, as in the latter physical, i.e. real-world, probability is considered.

Chapter 5

The Black and Scholes Framework and its Extensions

5.1. Black–Scholes Model — Part I

In 1973, Black and Scholes published a paper where it is discussed what is considered the most famous model in option pricing theory [8] and it represents a really simple and effective answer for the fair value estimation of a European call and put option. In the same year, Merton extended the understanding of the Black and Scholes mathematical framework, referring to the Black–Scholes (BS) paper as *Black–Scholes Options Pricing Model* [15].

In this section, we derive their formula in a very simple way, neglecting most of the formalism previously introduced about stochastic processes and fundamental theorems of the asset pricing. Actually, from a mathematical perspective, some steps in our derivation are not formally correct; anyway, these inaccuracies are justified by a simplicity of consideration and the final result is indeed correct. The main goal of this section is not to provide a formal derivation of the celebrated BS formula, but to give a good insight about its deep meaning, easing the intuition and the visualization of the main concepts. In the following sections, a more formal approach will be shown in order to assure coherence leveraging in our theoretical framework. In any case, we observe that, in the financial world, a simple, practical and intuitive explanation is often preferred to a formal and highly mathematical one; financial intuition should drive the mathematical modeling.

By the BS model, it is possible to obtain an analytical relation between the price of a fundamental underlying asset and the price of a contingent claim, assuming a lognormal dynamics for the underlying. In particular, by using BS model, it is possible to obtain an analytical formula for the pricing of a special class of financial derivatives: the *European call and put options*. As we will see in the following sections, the availability of simple analytical formulas for the pricing of financial instruments is very important as it allows one to induce the *unknown* parameter of the model just by comparing the model prices with the market prices. In fact, assuming that for liquid assets, the market price is equal to the fair value price, imposing the equivalence between the two prices, it is possible to *calibrate* the model parameter in order to obtain a good match between the model and the reality. We will come back to the model calibration in Section 5.4, here it is worthwhile to keep in mind that, as typically the calibration is quite a time-consuming procedure, the availability of analytical pricing formulas is quite compulsory in order to speed up the calculations.

The main hypotheses of the BS model are:

- It is possible to borrow and lend cash at a known constant risk-free interest rate.
- There are no transaction costs or taxes.
- The stock does not pay a dividend.
- All securities are perfectly divisible.
- There are no restrictions on short selling.
- There is no arbitrage opportunity.
- Options use the European exercise terms, which dictate that options may only be exercised on the day of expiration.
- The price of the underlying follows a lognormal Brownian motion:

$$dS(t) = \mu S(t)dt + \sigma S(t)dW(t), \tag{5.1}$$

where S is the price of the underlying, dt is the time increment, μ is a constant that represents the drift term, σ is the constant diffusion term and dW is the usual Wiener increment.

Starting from these hypotheses it is possible to build a fictitious portfolio composed by a long position of a call option (the derivation is similar for a put option) and a short position of a certain amount Δ of the underlying:

$$\Pi(S,t) = C(S,t) - \Delta(S,t)S(t). \tag{5.2}$$

Assuming that Δ does not change over the time interval dt, we can differentiate Eq. (5.2),

$$d\Pi(S,t) = dC(S,t) - \Delta(S,t)dS(t). \tag{5.3}$$

Applying Itô's lemma (Section 2.14) to the differential $dC(S, t)$, with dS given by Eq. (5.1), we find

$$dC = \frac{\partial C}{\partial t}dt + \frac{\partial C}{\partial S}dS + \frac{1}{2}\sigma^2 S^2 \frac{\partial^2 C}{\partial S^2}dt, \qquad (5.4)$$

and so, substituting the last equation in Eq. (5.3), we obtain

$$d\Pi(S, t) = \frac{\partial C}{\partial t}dt + \frac{\partial C}{\partial S}dS + \frac{1}{2}\sigma^2 S^2 \frac{\partial^2 C}{\partial S^2}dt - \Delta dS. \qquad (5.5)$$

From Eq. (5.5), it is clear that by choosing

$$\Delta = \frac{\partial C}{\partial S}, \qquad (5.6)$$

any source for randomness can be canceled making the theory insensitive to the random fluctuations of the market. Using the financial terminology, Δ is called the *delta* of the option and represents the amount of underlying that the writer of the option should buy to cancel the market risk (*delta-hedging*). From a mathematical point of view, this step is not fully consistent with the initial hypothesis that requires the Δ parameter to be constant. From a practical perspective, one can assume that $\frac{\partial C}{\partial S}$ is smooth enough so that it can be approximated by a constant value over a small time increment. This approach implies on the other hand that, in order to correctly hedge the position, one has to continuously update the delta estimation. In any case, since the stochastic term has been removed from the equation governing the evolution of Π, we impose

$$d\Pi(S, t) = \Pi(S, t)rdt, \qquad (5.7)$$

where r represents the risk-free interest rate, namely the rate of interest for an investment without risk. As a consequence, Eq. (5.5) becomes

$$\frac{\partial C}{\partial t} + \frac{\partial C}{\partial S}rS + \frac{1}{2}\sigma^2 S^2 \frac{\partial^2 C}{\partial S^2} = rC. \qquad (5.8)$$

This is a partial differential equation (PDE) that needs boundary conditions to be solved. This condition is given by the payoff of the call option at the time to maturity which is given by Eq. (3.32). Equation (5.8) can be solved using different techniques, for example, by Feynman–Kac theorem (Section 2.15.2), we can transform the PDE problem in an expectation estimation problem (Eq. (2.109)) that will be solved in Section 5.2

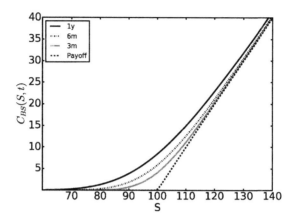

Fig. 5.1. We show the BS call option prices with strike price $K = 100$ as a function of the price of the underlying S. The different lines represent how the price changes for different fixed maturities. As evident, the shorter is the maturity, the higher is the similarity of the curve with the final call payoff.

(the explicit calculation is in Appendix A.1). The solution is given by (see Fig. 5.1)

$$C(S,t) = S(t)N(d_1) - Ke^{-r(T-t)}N(d_2), \tag{5.9}$$

where

$$d_1 = \frac{\ln(S(t)/K) + (r + \sigma^2/2)(T-t)}{\sigma\sqrt{(T-t)}},$$

$$d_2 = d_1 - \sigma\sqrt{(T-t)}, \tag{5.10}$$

$$N(x) = \frac{1}{\sqrt{2\pi}} \int_{-\infty}^{x} dz\, e^{-z^2/2}.$$

The structure of Eq. (5.9) is similar to the payoff (3.32), except for the fact that in this case, the spot price and the discounted strike price are weighted in a probabilistic manner by the cumulative of the Gaussian distribution, $N(x)$.

The solution of the BS model depends on five parameters: $K, T, r, S(t),$ and σ. Three of these five parameters can be easily deduced: the strike price (K) and the time to maturity (T) are written in the call contract and the spot price $(S(t))$ can be taken from the market. In the classical

BS framework, one typically assumes that the risk-free parameter (r) can be evaluated considering, for example, the London Interbank Offered Rate (LIBOR) for a fixed time to maturity. Actually, the choice of the suitable risk-free rate is not an easy task and other aspects such as liquidity risk, funding costs and collateral choices should be taken into consideration in order to define an appropriate risk-free rate (see Chapter 7). As this choice requires to take into consideration some technicalities that are out of the scope of this chapter, in the following, we will neglect all these aspects and we will simply assume that it is possible to define the risk-free rate in a proper way. As a consequence, the only remaining free parameter of the model is σ that represents the volatility of the underlying, i.e. the amplitude of the random (log) fluctuations of the underlying price and it has to be estimated by some calibration procedure that will be presented in Section 5.4.

This derivation of the BS formula is quite intuitive and gives us the opportunity to focus on some crucial points in the demonstration. First of all, one should observe that the final formula (see Eq. (5.9)) does not depend on the drift of the underlying μ. This is very a important aspect because at the very beginning of our demonstration, we actually *assumed* that the underlying follows a dynamics with a drift constant term μ (Eq. (5.1)). On the contrary, the price of the call option does not depend on this parameter, even if it is *contingent* to the underlying. This is quite a counter-intuitive result and it can be justified keeping in mind that we are just trying to *replicate* the call payoff with a strategy that involves the underlying, and this is possible as the value of the call is somehow related to the (contingent) value of its underlying. Given this relation, it is possible, by a suitable choice of Δ, to cancel the dependence of the drift term μ in the portfolio (made by the option and the underlying) together with the market fluctuations represented by the Wiener process. In other words, Eq. (5.6) tells us how to replicate call payoff in a dynamic and time-dependent way with two main achievements:

- the risk related to the market fluctuations is canceled out, and
- the drift term μ does not affect our estimation.

This is a very important result as it makes option valuation independent of future market expectations; in particular, under the replica assumption, the μ parameter can be substituted by the risk-free rate, implying, from a mathematical perspective, the need for a change of measure already discussed in the previous chapter.

5.2.　BS Model — Part II — Girsanov is Back

In this section, we are going to derive the famous BS applying the formalism of the risk-neutral pricing as discussed in Section 4.1. As one could expect, the final result of this second derivation is the same as the one obtained in Section 5.1, as a consequence, the same considerations expressed in the previous section still hold in this context. Here, we want to give a simple example of how the general theory of the risk-neutral pricing can be applied in order to derive pricing equations for simple option derivatives.

Following the algorithm described in Section 4.1.5, we firstly define the dynamics for the underlying. Assuming all the hypotheses in Section 5.1 hold, we look for a drift that assures that the discounted underlying is a martingale. This requirement leverages on the Girsanov theorem and implicitly defines the risk-neutral measure that has to be used for pricing purposes. Remembering the BS dynamics,

$$dS(t) = \mu S(t)dt + \sigma S dW(t), \tag{5.11}$$

we have that the discounted expected value (remember that the risk-free rate is assumed to be constant) under a generic measure $\mathbb{P}$ is given by

$$\mathbb{E}^{\mathbb{P}}[D(0,T)S(T)|\mathcal{F}(t)] = S(t)e^{(\mu-r)(T-t)}. \tag{5.12}$$

From this relation, it is clear that in order to obtain a martingale dynamics, we need to change the drift term from μ to r, finally obtaining the risk-neutral dynamics,

$$dS(t) = rS(t)dt + \sigma S(t)dW(t). \tag{5.13}$$

The next step is to express in mathematical terms the payoff of the financial instrument we want to price. In our case, we are mainly interested in a call where payoff is given by Eq. (3.32) and we report here for clarity:

$$C(S,T) = \max\left(S(T) - K, 0\right). \tag{5.14}$$

As a third step, all we need to do in order to obtain the price of a call option is to estimate the discounted expected value of the payoff under a risk-neutral measure,

$$C(S,t) = \mathbb{E}^{\mathbb{Q}}[e^{-(T-t)r}\max(S(T) - K, 0)|\mathcal{F}(t)]. \tag{5.15}$$

The solution of this equation can be obtained considering the definition of the expectation operator as an integral over a function of the probability density function (PDF) of the underlying that in this case is a lognormal distribution. The details about the derivation of the solution can be found in Appendix A.1. As one could expect, the result of this calculation is given in Eq. (5.9),

$$C(S,t) = S(t)N(d_1) - Ke^{-r(T-t)}N(d_2),$$ (5.16)

where $N(x)$, d_1 and d_2 are defined as in Section 5.1.

As a final step, we need to find a replica strategy for the call that allows us the leverage on no-arbitrage condition in order to price a financial instrument. From this point of view, we stress that if a concrete hedging strategy cannot be provided, any pricing formula becomes inconsistent with the theoretical framework and its correctness is not assured. In our risk-neutral pricing framework, assuming that all the hypotheses underlying it are satisfied, we have already demonstrated in Section 4.1.5 that the correct replica strategy is given by continuously exchanging a $\Delta(t)$ amount of underlying (and investing the remaining part in the bank account) that can be estimated by taking the derivative with respect to the underlying,

$$\Delta(t) = \frac{\partial C(S,t)}{\partial S(t)} = N(d_1).$$ (5.17)

In Appendix A.2, we report the analytical derivation of this solution.

As expected, we obtained the same results already mentioned in Section 5.1, applying the risk-neutral framework for a simple financial instrument.

5.3. Put–Call Parity

In the preceding sections, we learned how to price a contingent claim within the risk-neutral framework and we analyzed a concrete application for the pricing of a call option. Given our theoretical framework, the derivation of the BS formula for a put option is quite straightforward and we leave the reader its analytical derivation. In this section, we want to derive the analytical BS formula for a put option starting from a well-known model-free relation that is very useful in practical applications, the *put–call parity*.

The derivation of this formula can be obtained considering the graphical representation of different payoffs and trying to compose them in order to find a known payoff. So, the unknown price can be deduced using the replica approach. In our case, we consider the following self-financing strategy:

- go long on a call of maturity T and strike K,
- go short on a put of the same maturity and strike,
- go short on the underlying of the two options.

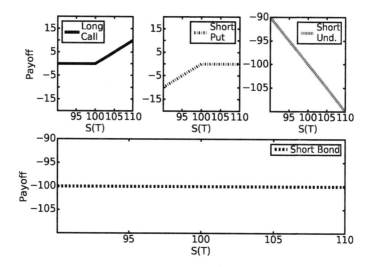

Fig. 5.2.   We show a graphical demonstration of the put–call parity. In particular, in the first line, we show the payoff of a long call, a short put and a short underlying position. As evident from the graphical analysis, it shows the results that the sum of the three payoffs gives the payoff of a bond as shown in the second line graph.

In Fig. 5.2, we show three payoffs at maturity T and their composition according to strategy. As evident from the picture, the net result of this strategy is constant line at $y = -K$ that is exactly the payoff that can be obtained going short on a bond of notional K. As in the usual replica framework, we can conclude that the proposed strategy is equivalent to a short position in a zero coupon bond (ZCB) so, in absence of arbitrage, the price of the whole portfolio must be equal to the bond price for all $t < T$. In formulas,

$$C(S,t) - \text{Put}(S,t) = S(t) - KP(t,T), \qquad (5.18)$$

where $P(t,T)$ is the price of the bond at time t and maturity T as described in Section 3.3.4.1. Equation (5.18) is known as *put–call parity* and it has a very fundamental role in the Quantitative Finance (QF) as it essentially shows that there exists an intuitive relation among nonlinear financial derivatives, the underlying instrument and a bond. In other words, this equation shows that in bond, stock and option markets, there must be a connection between them and a simple investment in a bond can be considered as a result of quite a complex investment strategy in options and stocks.

The put–call parity can be exploited in order to derive the price of a put given the price of the other instruments in the relation. As an example, we consider the BS price of a call option $C_{\mathrm{BS}}(S,t)$ given by Eq. (5.9). Because of the model's hypotheses, we are assuming that the risk-free rate is constant, so the ZCB price is simply given by Eq. (3.20). Under these assumptions, the put–call relation becomes

$$C_{\mathrm{BS}}(S,t) - \mathrm{Put}(S,t) = S(t) - Ke^{-r(T-t)}, \tag{5.19}$$

from which we can deduce the BS price of a put option. Using the property of the normal cumulative density function (CDF) $N(-x) = 1 - N(x)$, it is simple to prove that the final expression is given by

$$\mathrm{Put}_{\mathrm{BS}}(S,t) = N(-d_2)Ke^{-r(T-t)} - N(-d_1)S(t), \tag{5.20}$$

where d_1 and d_2 are defined as in Eq. (5.10). From the last equation, it is clear that the volatility related to the call and put options with the same strike and maturity must be the same.

This simple derivation provides an example of the strength of the put–call parity. What we want to stress here is that, as put–call parity (as described by Eq. (5.18)) was derived by simple replica and no-arbitrage principles, it is model independent and we were not required to make any explicit hypothesis on underlying dynamics. On the contrary, in order to derive the put price, we were required to make explicit the price of the call option that is model dependent; in this case, we used BS formula. For this reason, the price of the put relies on these assumptions and it is correct only if BS hypotheses hold.

5.4. Implied Volatility and the Calibration Problem

In the preceding sections, we derived analytical BS formulas for the pricing of call and put European options, starting from simple assumptions on the dynamics of the underlying. As already mentioned in Section 5.2, the call and put options are in general very liquid instruments, so their price can be obtained just by looking at the market price without the need for any mathematical model. At this point, a very natural question would be: *why is the BS model so important in the financial world and how is it used?*

The answer to this natural question is that BS model can be used to *infer* the future volatility of the market (assuming that all the BS assumptions hold) and we can also use it to price other (more complex) financial instruments using standard risk-neutral approach. In fact, as already observed in Section 5.1, BS model essentially depends on two parameters: σ that represents the *future* standard deviation of the log-returns and r which is the *future* risk-free interest rate. In general, people assume that r is a known parameter as it should be possible to proxy a bank account with the less risky investment available in the market that can provide/receive liquidity, receiving/paying an interest rate.[a] As a consequence, in the BS framework, the only free parameter of the model is assumed to be the future volatility. By "future volatility", we mean that we are interested in the future (average) fluctuations of the underlying in order to hedge them by delta (Eq. (5.17)) and to replicate the final payoff. From this point of view, past volatility, or historical volatility, could be used to this aim only if it would be perfectly constant in time so that the past estimation would be a suitable estimator for the future. Unfortunately, empirical studies show that, even if there is a sort of persistence that can hold for a few weeks, the empirical volatility is not constant in time as it can be observed in Fig. 5.3. In particular, it can be observed that

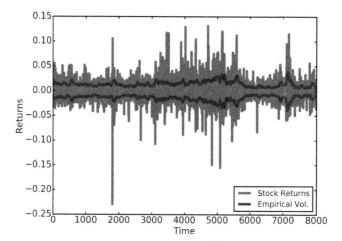

Fig. 5.3.   We show the historical series of relative price variation of a single stock time series and the empirical estimation of the (local) volatility (we represented it by double line symmetric with respect to the x-axis in order to better identify the typical range of the fluctuations). It is evident from the figure that the empirical volatility is not constant in time and it has large variations.

---

[a]Actually, finding a suitable bank account proxy is not an easy task; at the time of writing, a commonly accepted proxy for risk-free rate is EONIA in the EUR zone.

in some cases, the volatility can double with respect to its past values in a very short time period, leading to a completely wrong value of the call option.

These shortcomings should be interpreted as warnings for the use of the past volatility into the BS formula to price instruments. On the other side, recent studies [16–18] showed how to deal with these issues in order to obtain good performance, at least from the hedging perspective.

From the pricing perspective, the situation changes radically because, as already mentioned in Section 3.2, Quants are typically required to find models that are able to match market prices more than a theoretical fair value. From this point of view, even if you could find out the *exact* future volatility of an underlying and derive a *theoretical* call price different from the market price, it would be hard and, actually incorrect, to justify the use of this theoretical price to compute, for example, the P&L of a financial strategy involving that call. The reason is that, as the call is not exchanged at the evaluation time at the theoretical price, nobody sells (or buys) this call at that price, so you cannot close the position and actually realize that P&L.[b]

For this reason, from the pricing perspective, a very common approach is to *induce* the volatility parameter, relying on the so-called *calibration procedure* that consists in solving the inverse problem of finding the value of the volatility that when included in the BS formula would match the market price of the call/put options.

In formulas, one can look at the solution of the following *calibration problem*:

$$C_{\text{BS}}(\sigma_{\text{impl}}; r, S_0, T, K) = C_{\text{mkt}}(K, T), \tag{5.21}$$

where C_{BS} is the BS analytical formula as given by Eq. (5.9), C_{mkt} is the call market price that depends on the strike price K and the maturity T and σ_{impl} represents the unknown of our problem. As in the BS world, the only free parameter is the volatility σ, one can invert the formula by a simple numerical root finder and obtain a value of volatility for each couple K, T. The BS volatility obtained by this calibration algorithm is called *implied* volatility, σ_{impl}, and the function $\sigma_{\text{impl}}(K, T)$ is the *implied volatility*

---

[b]Obviously, you could always consider trading strategy in order to exploit this pricing mismatch and realize a positive P&L, but you cannot include this P&L straightforwardly in the balance sheet if you do not put the strategy in place.

surface. Typically, this surface is understood as the future expectation of the market participants of volatility.

Once the implied volatility surface is obtained, it is possible to use it to price financial instrument depending on the same underling but with unknown price, as all the parameters of the underlying dynamics are specified, i.e. r and σ, using, for example, risk-neutral formula, Eq. (4.18). Qualitatively speaking, as we are sure that all the known market prices are perfectly matched, we can expect that the unquoted price of an instrument can be coherently obtained using the same model parameters, following the principle that to find a price, we have to look at similar products, using a reasonable extrapolation, rule (see Section 3.2.1). In our case, the extrapolation rule is given by the implied volatility surface and the risk-neutral approach.

As we are in the BS framework, we observe that we cheated a little when we defined a volatility *surface* as one of the basic assumptions of the BS model is that the market volatility is constant. This generates quite a puzzling situation for the following reason: two call options on the same underlying with different strikes could have two different volatilities, so it is not clear which of the two volatilities represents the future volatility expectation of the underlying. In particular, it is not clear which volatility should be used to hedge the two products or a hypothetical third product that lies in between the two strikes. Obviously, the problem would not exist if the whole surface was perfectly flat. This would be the empirical proof that the constant volatility hypotheses of the BS model holds in practice and puzzling situations would be solved. Unfortunately, as one could expect from our previous discussion on historical volatility, the empirical implied volatility surfaces are not constant; on the contrary, they show a concavity with a minimum around the at-the-money strikes; typically, people refer to this concavity as *volatility smile*. This is the empirical confirmation that the BS model still requires some adjustment. In particular, the volatility smile is typically interpreted as a small perturbation to the flat volatility structure assumed by the model in order to take into account the fat-tails that characterize the empirical distributions of the financial returns. According to this interpretation, as extreme events occur with a higher probability than the ones that can be deduced by a Gaussian distribution, the implied volatility must be slightly increased in order to take into account this extreme events. This adjustment is required especially when the strike price is significantly different from the spot price, as in that case, the strike price is qualitatively in the region of this extreme events. This justifies why deep out-of-the-money and deep in-the-money options require a higher level of adjustment than at-the-money options.

Above, we referred to volatility smile as a small *perturbation* of the BS model. This expression should be understood as a warning in order to remember that the volatility is assumed to be constant in the BS framework; as a consequence, in order to keep the model coherent with the hypotheses, the variations in terms of volatility must be small and should be interpreted as perturbations.

It is possible to estimate how much we change the *implied distribution* of financial returns because of the introduction of a smile in the volatility structure. Starting from the risk-neutral relation, we have

$$C(S, K, r, T) = \mathbb{E}^{\mathbb{Q}}\left[e^{-rT}\max\left(S(T) - K, 0\right)\right]$$

$$= e^{-rT} \int_0^{+\infty} \max\left(S(T) - K, 0\right) p(S(T))\mathrm{d}S(T), \quad (5.22)$$

where we fixed time of the evaluation $t = 0$ and $p(S(T))$ is the PDF of prices at the maturity date T. We now try to isolate the PDF of the underlying prices, iteratively taking the derivative of this expression with respect to K. For the first derivative, we get

$$\frac{\partial C(K)}{\partial K} = -e^{-rT} \int_0^{+\infty} \mathbb{I}_{(K < S(T))} p(S(T))\mathrm{d}S(T)$$

$$= -e^{-rT} \int_K^{+\infty} p(S(T))\mathrm{d}S(T)$$

$$= -e^{-rT} \left(1 - \int_0^K p(S(T))\mathrm{d}S(T)\right), \quad (5.23)$$

where $\mathbb{I}_x$ is the indicator function defined in Chapter 2. Taking the second derivative, we get[19],

$$\frac{\partial^2 C(K)}{\partial K^2} = e^{-rT} p(K). \quad (5.24)$$

Equation (5.24) is a very general result that makes explicit the relation between the concavity of the price of a call option with respect to its strike and the implied PDF of prices of the underlying. In particular, in the BS framework, this equation highlights how, by introducing a smile effect in volatility surface, we change the implied distribution of prices because of the effect on the concavity of the call prices. In particular, it can be shown that if the perturbation in the volatility smile is strong enough, i.e. there is a significant deviation from the flat surface, the PDF of prices could become *negative*.

This effect is relevant not only from a theoretical perspective, but it also has practical implications: in particular, it can be shown that a negative PDF implies the existence of an arbitrage opportunity. The demonstration of this fact is quite simple if we consider a discrete approximation of the second derivative of the call price with respect to the strike:

$$\frac{\partial^2 C(K)}{\partial K^2} \sim \frac{C(K + \delta K) - 2C(K) + C(K - \delta K)}{\delta K^2}, \tag{5.25}$$

where δK is a small increment of the strike K. Equation (5.25) can be interpreted as a fictitious portfolio of three call options with strikes $K_1 = K - \delta K$, $K_2 = K$, $K_3 = K + \delta K$; in particular, we buy the first and the third option with strikes K_1 and K_3 and we sell a double amount of the call with strike K_2. Assuming the three options have the same maturity T, it is possible to obtain the final payoff of the whole portfolio, which is by definition always positive or equal to zero, as shown in Fig. 5.4.

Now, if at time $t < T$, the implied PDF would be negative, it means that it would be possible to implement a trading strategy that starts with a negative value and ends up with a positive or zero value, generating a positive P&L. As this strategy does not require any additional cash injection with the exception of the time zero, it is a self-financing strategy, so it could represent an arbitrage opportunity [20].

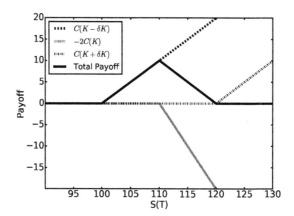

Fig. 5.4. We show three call payoffs described by Eq. (5.25) and the resulting total portfolio payoff. As expected, the total portfolio is always larger or equal to zero.

This simple example highlights the importance of the methodology for the construction of the implied volatility surface in the BS framework and its relevance in order to avoid the generation of arbitrage opportunities. In particular, it must be verified that the condition $\frac{\partial^2 C}{\partial K^2} \geq 0$ holds for all the surface implied by call prices.

In general, these kinds of problems arise whenever it is necessary to extrapolate information from market data (as in the case of the implied volatility). In a more general framework, one is interested in finding the set Σ of parameters that allows one to match a set of available prices with the best accuracy; we will refer to this task as *calibration problem*. In more formal terms, given

- a set of calibration instruments $I = \{1, \ldots, N\}$, where N is the total number of instruments,
- the market price of each instrument O_i^{mkt} $\forall i \in I$,
- a model dynamics MD,
- the pricing functions $O_i^{MD}(\Sigma)$ $\forall i \in I$ that depend on the set of parameters Σ and the specified model dynamics.

We want to solve the following calibration problem:

$$\min_{\Sigma} \|\omega_i(O_i^{mkt} - O_i^{MD}(\Sigma))\|^2, \tag{5.26}$$

where ω_i represents an array of weights and $\|\cdots\|^2$ is a suitable norm. The result of this calibration procedure is a set of parameters Σ that allows one to properly reproduce the prices of our set. In this general setup, the exact match of each price of the set is not assured and it depends on the number of parameters of the model and the number N of instruments used for the calibration. For example, in the BS framework, we could obtain the perfect matching of the prices only by introducing a different value of implied volatility for each instrument by that belongs to the surface. The choice of the required number of parameters to fit the prices of the calibration set depends on the model and the required degree of precision; in most of the cases, it is required to find a trade off between precision and numerical stability of the interpolation procedure.

5.5. Digital Options

In this section, we consider another example of application of the risk-neutral pricing approach to a common option exchanged in the market: the

Digital option. We have already discussed the payoff of this option in Section 3.4; in particular here, we consider the *cash-or-nothing call* payoff given by Eq. (3.34). In order to apply the risk-neutral approach, we consider the BS dynamics of the underlying given by Eq. (5.13) and we estimate the expectation of the payoff,

$$\text{Dig}(t, S_0) = \mathbb{E}^{\mathbb{Q}} \left[e^{-(T-t)r} \mathbb{I}\left(S(T) > K\right) | \mathcal{F}(t) \right], \tag{5.27}$$

where $\mathbb{I}(x)$ is the indicator function, introduced in Chapter 2, that summarizes the digital payoff. Rewriting this equation in integral form for time $t = 0$, we have

$$\text{Dig}(S_0) = e^{-rT} \frac{1}{\sqrt{2\pi\sigma^2 T}}$$

$$\times \int_K^{+\infty} \frac{1}{\sqrt{S(T)}} \exp\left[-\frac{\left(\log\frac{S(T)}{S_0} - \left(r - \frac{\sigma^2}{2}T\right)T\right)^2}{2\sigma^2 T} \right] dS(T). \tag{5.28}$$

We have already solved this integral in order to price the call option (see Appendix A.1 for details). As a result, the pricing function of the digital option is given, in the case of a cash-or-nothing call, by

$$\text{Dig}(t = 0, S_0) = \text{Dig}(S_0) = e^{-rT} N(d_2), \tag{5.29}$$

where $N(x)$ is the usual Normal CDF and d_2 is defined as in Eq. (5.10). This result provides us a useful interpretation of $N(d_2)$ as appearing in the BS formula as the probability of the underlying of being higher than the strike price. In addition, this result suggests that it should be possible to find a relation between call and put options and the digital option. Actually, taking the put and call options as fundamental instruments, one could try to reproduce the digital payoff in order to price this financial instrument with a replica approach. For example, one can consider the following strategy:

- buy a $1/\delta$ amount of a call with strike K and maturity T, and
- sell a $1/\delta$ amount of a call with strike $K + \delta K$ and the same maturity T.

At maturity, the resulting payoff is something very similar to the digital payoff shown in Fig. 5.5 with the exception of the region between K and $K + \delta K$.

This kind of payoff is called *bull call spread*. It should be quite intuitive that as δ goes to 0, the region of the payoff between K and $K + \delta$ decreases,

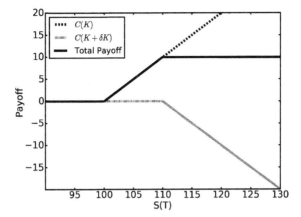

Fig. 5.5. We show the two call payoffs considered for the approximation of the digital option payoff. The straight line represents the resulting total portfolio.

approximating the digital option payoff. In order to derive the digital option pricing formula (holding BS hypothesis), we need to calculate the following limit:

$$\text{Dig}(S_0) = \lim_{\delta \to 0} \frac{S_0 N(d_1) - Ke^{-rT} N(d_2) - S_0 N(d_1^\delta)}{\delta}$$
$$+ \frac{Ke^{-rT} N(d_2^\delta) + \delta e^{-rT} N(d_2)}{\delta}, \tag{5.30}$$

where we made use of the BS formula, Eq. (5.9), and we used the notation d_j^δ to refer to the use of $K + \delta K$ strike in the equations for d_1, d_2. This limit can be decomposed in two parts; the first part can be proved to be equal to 0 by applying De L'Hopital theorem and making use of Eq. (8.25),

$$\lim_{\delta \to 0} \frac{S_0 N(d_1) - Ke^{-rT} N(d_2) - S_0 N(d_1^\delta) + Ke^{-rT} N(d_2^\delta)}{\delta} = 0. \tag{5.31}$$

As a consequence, we obtain for $\text{Dig}(S_0)$ the same result of Eq. (5.29) assuring that our replica and risk-neutral approach are consistent. In this case, we stress that in our replica, we made use of the BS expression for call option in order to obtain Eq. (5.29). For this reason, the replica result is in this case *model dependent*.

With this simple example, we showed how it is possible to apply the replica approach to nonlinear payoffs, including in the set of our fundamental products also call and put options. From a very theoretical point of view, this would suggest that the market of linear products is not complete, i.e. we cannot reproduce the price of any contingent claim only using fundamental instruments; for this reason, we are forced to introduce in our set of fundamental instruments some nonlinear payoffs like call and put options. This idea is in contrast with our risk-neutral pricing framework, where we could replicate call and put options with the delta-hedging of the underlying, implying that calls and puts market is somehow redundant from this perspective. Probably, given the huge number of transactions related to call and put options, it is unlikely that this market is redundant and could be simply substituted by the delta-trading of the underlying. In our opinion, this is a consequence of the impossibility of hedging all the risks by the use of the underlying, but additional nonlinear instruments are required. We will further discuss this aspect in Section 5.8.

5.6.   Change of Numeraire

In Section 3.3.1, we dealt with the concept of bank account and discount factor and we observed that they represent a way to propagate backward and forward in time the value of money. In practice, we divide the value of our instrument by the bank account (or equivalently, we multiply it by the discount factor) to understand how much our investment is profitable *with respect to* the bank account. In other words, in order to propagate through time the value of an investment, we *compare* its value with a reference instrument that embeds the standard risk-free gain. The ratio of the two investments represents a very fundamental quantity as its expected value gives us the price of the instrument, Eq. (4.18).

Actually, one can generalize this approach, considering a generic reference instrument, called *numeraire* and estimating the value of other instruments with respect to it. It can be proved that, if suitable requirements are satisfied, the pricing formula, given by (Eq. (4.18)), can be generalized with respect to a generic numeraire and under a suitable probability measure that can be derived, exploiting the change of measure relations about the conditional expectation described in Section 2.15.3. The importance of this generalization is twofold:

- From a theoretical point of view, it shows that after all the bank account is not a special instrument and a pricing algorithm can be obtained just by considering the excess of performance of an instrument with respect to *a* reference instrument with suitable characteristics.

- From a practical point of view, we have some degrees of freedom in the choice of the measure, or, equivalently, of the numeraire, to be used in the pricing algorithm. This fact can be exploited in order to simplify the calculations.

In this section, we derive a generalization of the pricing formula Eq. (4.18), finding the Radon–Nikodym derivative (Section 2.15.3) for a generic numeraire and applying the change of measure technique. We think that this generalization has very important implication from a theoretical point of view as it clearly shows the role of the numeraire in the pricing equation and this is the reason why we decided to include this rather technical argument in an introductive book on QF.

From a practical point of view, we show an application of this formula for pricing a forward.

To start with our derivation, we firstly define the numeraire as a *positive paying tradable asset that does not pay dividends*. This definition is compatible with our first numeraire choice, i.e. the bank account. In particular, we observe that we introduced it as a fundamental instrument that can be used to receive a risk-free interest rate or to borrow money, so it is a tradable asset. In addition, it is positively defined because of the exponential function in its definition (Eq. (3.2)). From this definition, assuming that the numeraire at time t is represented by the random variable $Z(t)$, we know that its value is given by the risk-neutral pricing formula, i.e.

$$\mathbb{E}^{\mathbb{Q}}\left(\frac{Z(T)}{B(t,T)}\bigg|\mathcal{F}_u\right) = \frac{Z(u)}{B(t,u)}, \tag{5.32}$$

where $\mathbb{Q}$ is the risk-neutral measure and B is the bank account. Equation (5.32) holds as a consequence of the fact that $Z(t)$ is a tradable asset, so it can be priced by usual pricing formula. In particular, the numeraire rebased $Z(t)$ is a martingale, so the martingale representation theorem (Eq. (2.102)) holds:

$$\mathrm{d}(D(0,t)Z(t)) = \phi(t)\mathrm{d}W(t), \tag{5.33}$$

where $\phi(t)$ is a suitable representation function and $W(t)$ is defined under the risk-neutral probability measure. Equation (5.33) describes how the numeraire $Z(t)$ evolves under the risk-neutral measure, given $\phi(t)$. Actually, we can define many numeraires and obtain many new measure definitions, exploiting the arbitrariness of $Z(t)$ or equivalently its representation $\phi(t)$ under $\mathbb{Q}$.

In the following, we fix

$$\phi(t) = \zeta(t)D(0,t)Z(t), \tag{5.34}$$

where $\zeta(t)$ is a positive enough smooth function, and we find the Radon–Nikodym derivative that implicitly defines the new numeraire measure as a function of the risk-neutral measure $\mathbb{Q}$ and the function $\zeta(t)$. In order to do this, we substitute Eq. (5.34) into Eq. (5.33)

$$\mathrm{d}(D(0,t)Z(t)) = \zeta(t)D(0,t)Z(t)\mathrm{d}W(t). \tag{5.35}$$

We now consider the function $f(Y) = Z(0)\exp(Y(t))$, where we defined

$$\mathrm{d}Y(t) = \zeta(t)\mathrm{d}W(t) - \zeta(t)^2\mathrm{d}t, \tag{5.36}$$

and we apply the Itô's lemma,

$$\mathrm{d}f = Z(0)e^{Y(t)}\mathrm{d}Y(t) + Z(0)e^{Y(t)}\zeta^2(t)\mathrm{d}t. \tag{5.37}$$

Substituting Eq. (5.36), we obtain

$$\mathrm{d}(Z(0)e^{Y(t)}) = Z(0)e^{Y(t)}\zeta(t)\mathrm{d}W(t) - Z(0)e^{Y(t)}\zeta(t)^2\mathrm{d}t$$
$$+ Z(0)e^{Y(t)}\zeta^2(t)\mathrm{d}t$$
$$= Z(0)e^{Y(t)}\zeta(t)\mathrm{d}W(t). \tag{5.38}$$

The last expression is equal to the right-hand side of Eq. (5.35) with the same initial conditions, i.e. $f(Y(0)) = Z(0) = D(0,0)Z(0)$, so $f(Y(t))$ *solves* Eq. (5.35),

$$D(0,t)Z(t) = Z(0)\exp\left(\int_0^t \zeta(u)\mathrm{d}u - \int_0^t \zeta^2(u)\mathrm{d}u\right), \tag{5.39}$$

where we made explicit the definition of the stochastic differential to express $Y(t)$. Using the discount factor definition Eq. (3.8), we finally obtain

$$Z(t) = Z(0)\exp\left(\int_0^t \zeta(u)\mathrm{d}W(u) - \int_0^t (\zeta^2(u) - r(u))\mathrm{d}u\right), \tag{5.40}$$

where $r(t)$ is the risk-free rate.

Comparing Eq. (5.39) with the Girsanov expression of the change of measure (Eq. (2.124)), we obtain the Radon–Nikodym derivative,

$$\frac{\mathrm{d}\mathbb{P}_Z}{\mathrm{d}\mathbb{Q}}(\omega) = \frac{D(0,t)Z(t)}{Z(0)}, \tag{5.41}$$

where $\mathbb{P}_Z$ represents the new probability measure as implied by the numeraire $Z(t)$. In addition, comparing Eq. (5.40) with Eq. (2.98), we can obtain the Stochastic Differential Equations (SDEs) for the dynamics of the numeraire $Z(t)$ under the risk-neutral probability,

$$dZ(t) = r(t)Z(t)dt + Z(t)\zeta(t)dW(t). \tag{5.42}$$

The last equation provides a clear insight about the role of the function $\zeta(t)$ we introduced above in order to control the martingale representation function. In fact, Eq. (5.42) implies that $\zeta(t)$ represents *the volatility of the numeraire.*

We can summarize the results we obtained till now:

- We observed that the bank account is just a way to "normalize" the performance of an asset with respect to a benchmark, so it should be possible to generalize risk-neutral pricing formula with respect to a generic benchmark.
- We defined the numeraire as a non-dividend paying tradable asset and we observed that its discounted value is a martingale in the risk-neutral measure.
- We modeled the numeraire dynamics by the volatility function $\zeta(t)$ and we obtained its Radon–Nikodym derivative to switch from the risk-neutral measure to the new numeraire measure (Eq. (5.41)).

Given these results, the last step to obtain our generalization is to apply Eq. (2.116) using the Radon–Nikodym derivative (Eq. (5.41)). In order to do this, we need to estimate the following expectations:

$$\mathbb{E}^Q \left(\frac{D(0,T)Z(T)}{Z(0)} \frac{V(T)}{Z(T)} \bigg| \mathcal{F}_t \right) = \frac{D(0,t)}{Z(0)} \mathbb{E}^Q \left(D(t,T)V(T) | \mathcal{F}_t \right), \tag{5.43}$$

where $V(T)$ represents the generic payoff at time T and we exploited the definition of the discount factor in order to split it at time t. Analogously, we obtain

$$\mathbb{E}^Q \left(\frac{D(0,T)Z(T)}{Z(0)} \bigg| \mathcal{F}_t \right) = \frac{D(0,t)}{Z(0)} \mathbb{E}^Q \left(D(t,T)Z(T) | \mathcal{F}_t \right)$$

$$= \frac{D(0,t)}{Z(0)} Z(t), \tag{5.44}$$

where in the last passage, we used the fact that the discounted numeraire is a martingale under the risk-neutral measure.

Now, applying Eq. (2.116), we obtain

$$\mathbb{E}^{\mathbb{P}_Z}\left(\frac{V(T)}{Z(T)}\bigg|\mathcal{F}_t\right) = \frac{\mathbb{E}^{\mathbb{Q}}\left(D(t,T)V(T)|\mathcal{F}_t\right)}{Z(t)}, \tag{5.45}$$

or reverting the equation,

$$\mathbb{E}^{\mathbb{Q}}\left(D(t,T)V(T)|\mathcal{F}_t\right) = Z(t)\mathbb{E}^{\mathbb{P}_Z}\left(\frac{V(T)}{Z(T)}\bigg|\mathcal{F}_t\right). \tag{5.46}$$

In the left-hand side, of the last equation, we have the risk-neutral pricing formula and on the right-hand side, its alternative version, expressed in the new measure $\mathbb{P}_Z$ implied by the numeraire $Z(t)$. In other words, Eq. (5.46) represents the generalization of the risk-neutral pricing formula for a generic numeraire. In addition, recollecting that the expected value of the discounted payoff is a martingale under the risk-neutral probability, Eq. (5.46), it can be proved that

$$\mathbb{E}^{\mathbb{P}_Z}\left(\frac{V(T)}{Z(T)}\bigg|\mathcal{F}_t\right) = \frac{V(t)}{Z(t)}. \tag{5.47}$$

That is, the expected value of the payoff divided by a numeraire is a martingale under the related numeraire measure. This property holds for all tradable assets and attainable claims.

As already mentioned above, the theoretical importance of Eq. (5.46) is to show the actual role of the discount factor (or equivalently of the bank account) in the risk-neutral pricing formula and how it is possible to extend it to a generic tradable (non-dividend paying) asset. From a practical point of view, Eq. (5.46) can be exploited to simplify calculations or price estimations as will be shown as an example in the next section.

5.6.1. *The Forward Measure*

The change-of-measure technique is particularly useful when we want to price an interest rate (IR)-linked product as the correlation between the risk-free rate and other interest rates in the payoff cannot be neglected. In this case, one cannot assume that the discount factor can be factorized outside the expected value of the risk-neutral pricing formula (as done in the BS case) and, as a consequence, the expectation calculation can be a difficult task. In this situation, one can exploit a change of measure in order to simplify the problem.

As an example, we consider the FRA par-value as in Eq. (3.31). In this case, we defined the par-rate as the rate that allows the FRA contract to be zero at inception. Here, we want to exploit Eq. (3.31) to *define* the *forward rate* as

$$F(t, S, T) \equiv \frac{1}{T - S}\left(\frac{P(t, S)}{P(t, T)} - 1\right), \tag{5.48}$$

where $t \leq S < T$. In this context, $F(t, S, T)$ is the rate implied by the two bonds $P(t, T)$ and $P(t, S)$ that describe the rate fixed at time S and paid at time T as seen at time t. In principle, $F(t, S, T)$ has nothing to do with the FRA rate, but we showed in Section 3.3.4.4 that, *under suitable hypotheses*, the forward rate is equal to the FRA par-rate.

The forward rate has a fundamental role in interest-rate modeling as it is the building block of many interest-rate derivatives; as a consequence, the modeling of its dynamics represents the first step in order to price IR derivatives, exactly as the modeling of the underlying equity asset is required to price call and put equity options, or more complex equity derivatives. For these reasons, we need to estimate the expected value of a forward rate under a suitable measure; in particular, we are interested in finding a measure definition that assures the forward rate to be a martingale, so that its dynamics can be simplified and one can focus only on the diffusion part,

$$\mathrm{d}F(t) = \sigma(F, t)\mathrm{d}W(t). \tag{5.49}$$

In order to do this, we consider the ZCB $P(t, T)$ and we observe that it satisfies all the requirements to be a numeraire, i.e. it is a (no-dividend paying) tradable asset and has non-negative value. As a consequence, we can apply Eq. (5.46) for the generic payoff $V(T)$ (remember that $P(T, T) = 1$),

$$V(t) = P(t, T)\mathbb{E}^{\mathbb{Q}_{P(t, T)}}\left(V(T)|\mathcal{F}_t\right), \tag{5.50}$$

where $\mathbb{E}^{\mathbb{Q}_{P(t, T)}}$ is the expected value under the $\mathbb{Q}_{P(t, T)}$-measure implied by the $P(t, T)$-numeraire.

Now, we want to estimate the expected value of the forward, under the measure defined above. In order to do this, we observe that

$$F(t, S, T)P(t, T) = \frac{1}{T - S}\left(\frac{P(t, S)}{P(t, T)} - 1\right)P(t, T)$$

$$= \frac{P(t, S) - P(t, T)}{T - S} \tag{5.51}$$

that implies that the product $F(t, S, T)P(t, T)$ is a tradable asset as it is a quantity proportional to the difference of two bonds (see right-hand side of the equation). As a consequence of Eq. (5.47), this quantity is a martingale for the chosen numeraire measure. This implies that

$$\mathbb{E}^{\mathbb{Q}_{P(t,T)}}\left(F(t, S, T)|\mathcal{F}_u\right) = \mathbb{E}^{\mathbb{Q}_{P(t,T)}}\left(\frac{F(t, S, T)P(T, T)}{P(T, T)}\bigg|\mathcal{F}_u\right)$$

$$= \frac{F(u, S, T)P(u, T)}{P(u, T)}$$

$$= F(u, S, T), \tag{5.52}$$

or in other words, that the forward rate is a martingale under the $\mathbb{Q}_{P(t,T)}$ measure which for this reason is called T-*forward* measure. This result is of practical importance as it allows one to postulate simple dynamics for the forward rate (see Eq. (5.49)) whenever the T-forward measure is used instead of the risk-neutral measure. In particular, it is not necessary to specify the drift term. The result of Eq. (5.52) holds only if the chosen numeraire $P(t, T)$ and the $P(t, T)$-bond in Eq. (5.48) are the same. Unfortunately, after the credit crunch of 2007, this is not the case and same modifications to the theoretical framework are required. In Section 7.2, we will discuss this case in depth and we will provide some simple solutions to this problem.

5.7. Greeks

In Sections 5.1 and 5.2, we derived the BS equation, imposing that the number of shares of the underlying $S(t)$ equals the derivative of the call price with respect to $S(t)$ (Eq. (5.6)). In both sections, we justified this choice in order to be able to replicate the contingent claim and to cancel out the risk related to the market fluctuations due to the terms dW (see Eqs. (4.12) and (4.16) in Section 4.1.5 and Eq. (5.5) in Section 5.1). We have already referred to this approach as a hedging procedure to cancel risk. Actually, it can be proved that this approach can be generalized to any (Brownian) source of risk in the context of risk-neutral pricing. In particular, in order to calculate the risk-neutral price, we are required to hedge risks, by buying or selling a given amount of different underlyings, ideally one underlying for each source of risk. In this section, we will generalize the risk-neutral approach shown in Section 4.1 to many sources of noise and we will prove that in this case also hedging can be obtained considering the derivatives of the contingent claim price with respect to each underlying. We stress from the very beginning that, in order to cancel out the noise, i.e. the risk, we should be able to buy or sell this noise in the market, or, in other words, the noise should be effectively represented by a tradable asset that can incorporate this risk. If this were not the case, the financial product could not be completely hedged. As a consequence, even if a hedging strategy was considered, the product would be risky and the market could not be considered complete with respect to this product (see Section 5.8).

In order to generalize our hedging approach to many sources of risk, we rely on the derivation described in Section 4.1, and in particular in Sections 4.1.4 and 4.1.5. In this case, we will give just a sketch of the proof in order to focus on the general framework and will not involve ourselves in technicalities. In line with Section 4.1.2, we need to define an appropriate dynamics; in this case, we assume that we have m underlyings $S(t) = (S_1(t), \ldots, S_m(t))^\dagger$ and d sources of randomness $W(t) = (W_1(t), \ldots, W_d(t))^\dagger$ that we want to hedge. In this context, we made use of the symbol $\dagger$ in order to indicate the transposition operator. As a consequence, S and W are column vectors of dimension $[m \times 1]$ and $[d \times 1]$, respectively. As a result, for the ith underlying, we postulate the following dynamics:

$$\mathrm{d}S_i(t) = \mu_i(t)S_i \mathrm{d}t + S_i \sum_{j=1}^{d} \sigma_{ij}(t)\mathrm{d}W_j, \qquad (5.53)$$

where $\mu_i(t)$ and $\sigma_{ij}(t)$ are the drift and diffusion parameters of the process, represented respectively by an m-dimensional vector and a $[m \times d]$-dimensional matrix. In addition, we define in the usual way the discount factor as in Eq. (3.8), where $r(t)$ is the risk-free rate. In particular, we use the shorthand notation $D(t) \equiv D(u, t)$. It is possible to prove that if $\Theta_1(t), \ldots, \Theta_d(t)$ are unknown processes and if the following equation, known as *market price of risk equations*, holds,

$$\mu_i(t) - r(t) = \sum_{j=1}^{d} \sigma_{ij}\Theta_j(t), \qquad (5.54)$$

using a generalized formulation of the Girsanov theorem, it is possible to transform the discounted dynamics of Eq. (5.53) in an SDE satisfying the martingale property,

$$d(D(t)S_i(t)) = D(t)S_i(t) \sum_{j=1}^{d} \sigma_{ij} dW^{\mathbb{Q}}(t), \tag{5.55}$$

where $dW^{\mathbb{Q}}(t)$ is the Wiener increment in the usual risk-neutral measure. In the following, we will not specify the risk-neutral measure in order to simplify the notation.

Once we obtained the desired martingale dynamics for the discounted underlying, we need to apply the bidimensional version of Itô's lemma, Eq. (2.101), in order to derive the discounted dynamics for a contingent claim price described by the generic function $V(S,t)$:

$$D(t)V(S,t) = (\ldots) dt + D(t) \left(\nabla_S V(S,t) \right)^{\dagger} G(S,t) dW(t), \tag{5.56}$$

where $\nabla_S V(S,t)$ is the gradient of $V(S,t)$ with respect to S and $G(S,t)$ is a matrix defined as

$$G(S,t) = \begin{pmatrix} S_1(t)\sigma_{11} & \cdots & S_1(t)\sigma_{1d} \\ S_2(t)\sigma_{21} & \cdots & S_2(t)\sigma_{2d} \\ \vdots & \ddots & \vdots \\ S_m(t)\sigma_{m1} & \cdots & S_m(t)\sigma_{md} \end{pmatrix}. \tag{5.57}$$

In this case, as we know that the discounted underlyings are martingales, we can get rid of the drift part of Eq. (5.56) and focus on the diffusion part. In other words, we need to perform the following calculation:

$$\left(\nabla_S V(S,t) \right)^{\dagger} G(S,t) dW(t) = \begin{pmatrix} \dfrac{\partial V(S,t)}{\partial S_1(t)} & \cdots & \dfrac{\partial V(S,t)}{\partial S_m(t)} \end{pmatrix}$$

$$\cdot \begin{pmatrix} S_1(t)\sigma_{11} & \cdots & S_1(t)\sigma_{1d} \\ S_2(t)\sigma_{21} & \cdots & S_2(t)\sigma_{2d} \\ \vdots & \ddots & \vdots \\ S_m(t)\sigma_{m1} & \cdots & S_m(t)\sigma_{md} \end{pmatrix}$$

$$\begin{pmatrix} dW_1(t) \\ \vdots \\ dW_d(t) \end{pmatrix} \tag{5.58}$$

that can be written in the following form:

$$dV(S,t) = \sum_{j=1}^{d}\sum_{i=1}^{m} \frac{\partial V(S,t)}{\partial S_i(t)} S_i(t)\sigma_{ij} dW_j^{\mathbb{Q}}(t), \qquad (5.59)$$

where we remark again that this result holds because of our choice of the risk-neutral measure. It can be observed that Eq. (5.59) represents a multi-dimensional generalization of Eq. (4.16).

As a third step, we need to define a suitable hedging portfolio. In analogy with Eq. (4.9), we define

$$dX(t) = \sum_{i=1}^{m} \Delta_i(t)dS_i(t) + r(t)\left(X(t) - \sum_{i=1}^{m} \Delta_i(t)S_i(t)\right)dt, \qquad (5.60)$$

where we define $\Delta(t) = (\Delta_1(t),\ldots,\Delta_m(t))^{\dagger}$ as a generalization of the Δ function considered in the mono-dimensional case. Following a similar derivation, as outlined in Section 4.1.4 and using martingale property of the discounted underlying, it can be proved that

$$d(D(t)X(t)) = \sum_{j=1}^{d}\sum_{i=1}^{m} \Delta_i(t)S_i(t)\sigma_i j(t)dW_j^{\mathbb{Q}}(t). \qquad (5.61)$$

As a final step, comparing Eq. (5.61) with Eq. (5.59), we obtain that in order to hedge any risk in our portfolio, we need to impose that

$$\Delta_i(t) = \frac{\partial V(S,t)}{\partial S_i(t)}, \qquad (5.62)$$

which represents a generalization of Eq. (4.17).

As already discussed above, this result has the intuitive interpretation that if we want to cancel out different sources of risk, we need to

- identify them,
- find a tradable asset that can represent one or more sources of randomness,
- estimate how much the price of our contingent claim depends on each source of risk (this is exactly the meaning of partial derivatives), and
- build a suitable portfolio able to hedge any source of risk.

As already outlined, the most difficult part is to find a tradable representation of each risk that would imply the completeness of the market.

In practice, the general idea that the partial derivative of the contingent claim price with respect to a generic underlying can be used to replicate a financial product and hedge the noisy part of the underlying dynamics is somehow extended to any kind of mathematical derivatives and any source of risk. In particular, the partial derivative is interpreted as a way to determine the degree of dependency between the product of interest and the considered risk factor and, for this reason, it is generically called *sensitivity* to the risk factor. Once you determine the sensitivity to each risk-factors, you can try to cancel all the risks by hedging them using the related sensitivity. From a mathematical perspective, this *sensitivity approach* is related to the Taylor expansion,

$$dO(t) = \sum_{i=1}^{N} \frac{\partial O(t)}{\partial S_i(t)} dS_i(t) + \frac{1}{2} \sum_{i=1}^{N} \frac{\partial^2 O(t)}{\partial S_i^2(t)} (dS_i(t))^2 + \cdots , \qquad (5.63)$$

where $O(t)$ is the price of the contingent claim, $S_i(t)$ is a generic risk factor and N is the total number of the risk-factors we consider. In this case, $S_i(t)$ assumes a wider meaning than in the context of the risk-neutral pricing, as it represents any source of risk and not only the underlying instrument. For example, in the BS framework, one could be interested in estimating the sensitivity to the implied volatility and try to hedge it, even if this is not required for the replica strategy, under the BS hypothesis.

From a theoretical point of view, this approach is quite different and it cannot be justified in the risk-neutral framework. Informally speaking, if we have a generic portfolio $\Pi(t)$ composed by a contingent claim $O(t)$ and different amount Δ_i of hedging instruments $S_i(t)$,

$$\Pi(t) = O(t) - \sum_{i=1}^{N} \Delta_i(t) S_i(t) \qquad (5.64)$$

in the *risk-neutral* framework, we want to hedge only the dependence of the price on the noisy part, i.e.

$$\frac{\partial \Pi(t)}{\partial W_i(t)} = 0 \qquad \forall i = 1, \ldots, N. \qquad (5.65)$$

This equation is obviously wrong from a mathematical perspective as the derivative with respect to a Wiener process is not defined. Equation (5.65) should be understood as the following generic rule: "take what is in front of the differential dW and equal it to zero".

On the contrary, in the *sensitivity framework*, we require (to the first order approximation)

$$\frac{\partial \Pi(t)}{\partial S_i(t)} = 0 \qquad \forall i = 1, \ldots, N, \tag{5.66}$$

where the partial derivative should be interpreted as a sensitivity. In both cases, we already know that the solution of the hedging problem is the same, i.e. $\Delta_i(t) = \frac{\partial O(t)}{\partial S_i(t)}$, proving that the two approaches are equivalent to the first order at least when dealing with $S_i(t)$ tradable assets, i.e. when a risk-neutral representation of the risk factor is possible. The reason for this equivalence is that in the risk-neutral approach, we are requiring that each discounted $S_i(t)$ is a martingale, so the differential of dS depends only on dW. On the contrary, when a risk-neutral representation is not allowed for all risk factors, the two approaches can be different, requiring different hedging strategies. For example, in the sensitivity framework, the hedging of implied volatility is required (and typically performed by the use of vanilla options) while it is not expected in the risk-neutral framework.

In practice, some sensitivity covers a very important role and it is of big interest in hedging activities. For this reason, these sensitivities have a specific name derived from Greek letters. In the following table, we summarize the most important ones.

Name	Formula	Description
Delta	$\Delta = \dfrac{\partial O(t)}{\partial S(t)}$	Sensitivity to the underlying. In our risk-neutral framework, delta hedging should be done continuously in time, but this is not possible for all practical purposes because of transaction costs. For this reason, a formulation of the problem in terms of optimization of the costs is typically required.
Gamma	$\Gamma = \dfrac{\partial^2 O(t)}{\partial S(t)^2} = \dfrac{\partial \Delta(t)}{\partial S(t)}$	Sensitivity of the delta of the option. In general, the estimation of a second-order derivative is not an easy task and this typically causes numerical instability.
Vega	$\upsilon = \dfrac{\partial O(t)}{\partial \sigma(t)}$	Sensitivity to the implied volatility of the option. In order to hedge this sensitivity, a proxy is required as it is not directly observable in the market. Typically, the hedge is obtained considering combination of call and put options or a special instrument called variance swap.

Name	Formula	Description
Rho	$\rho = \dfrac{\partial O(t)}{\partial r(t)}$	Sensitivity to the risk-free rate. The hedging of this sensitivity is not straightforward as the risk-free rate is typically a function of time and the maturity considered. Linear financial instruments are typically used to hedge this risk.

5.8. Heston Model

5.8.1. *Incomplete Markets*

In Sections 5.1 and 5.7, we showed how it is possible to cancel the market fluctuations of a call option, buying a suitable amount (the greek) of the underlying or the related risk factor. In order to ensure that the approach is valid, we essentially required that

- The dynamics of the risk factor can be represented by a continuous process in order to assure that the martingale representation theorem can be applied.
- It is possible to trade the risk factor in the market or a closely related financial instrument with a high correlation.

In particular, if the second hypothesis holds, we can say that the market is *complete* as all sources of risk can be hedged with the greeks approach.

In this section, we deal with models where market completeness is not postulated and some source of risk cannot be directly hedged. This fact has strong consequences on our theoretical framework as an exact replica of all the financial products cannot be found. As a consequence, the price is not unique and it will depend on the *premium of risk* that the market participants will require in order to take the risk that cannot be hedged. In mathematical terms, this implies that there is not a unique risk-free measure but a set of equivalent martingale measures that can be arbitrarily chosen as a function of the required risk premium. If, at first sight, this could be understood as a model deficiency, this feature can actually make the model more realistic. In fact, the market completeness hypothesis is equivalent to assume that derivatives market is completely redundant as it can be exactly replicated by suitable strategies based on underlyings. On the contrary, the derivatives market is very active and in practical situations, and an exact replica of a derivative can be hardly achieved. This empirical evidence confirms that market completeness should be regarded as an interesting way to *approximate* reality, which is actually much more complex. On the other side, as there is no unique equivalent martingale measure, the model calibration will need to deal with the estimation of the risk premium.

5.8.2. *Heston Model Equations*

In this section, we are going to describe the so-called *Heston model* that was derived by S. L. Heston in 1993 [21]. The Heston model assumes that the volatility of the underlying changes randomly, following a given stochastic dynamics, in order to take into consideration the volatility smile effect (Section 5.4). In particular, the Heston model allows for an increase in the probability of extreme events since volatility randomly fluctuates around a (time-dependent) mean level; in this way, it is possible to explain high prices for not at-the-money call options as implied by volatility smile effect. As the volatility of a financial instrument is not a traded asset,[c] volatility fluctuations cannot be hedged; as a consequence, this model makes the market incomplete.

In order to model the underlying dynamics, Heston model assumes the following SDE:

$$dS(t) = \mu S(t)dt + \sqrt{\nu(t)}S(t)dW_1(t),$$

$$d\nu(t) = k\left(\theta - \nu(t)\right)dt + \eta\sqrt{\nu(t)}dW_2(t), \tag{5.67}$$

where μ is the drift of the underlying in the physical measure $\mathbb{P}$, $\nu(t)$ is the stochastic variance of the underlying described by the second SDE in Eq. (5.67) with parameters k, θ, η. Looking into more detail at the latter equation, we can observe that, if we neglect the diffusion part of the equation that is purely random, the sign of the squared volatility variation $d\nu(t)$ depends on the level of $\nu(t)$ with respect to a reference level given by θ. As a consequence, on average, when $\nu(t)$ is less than θ, we will observe a positive variation, while we will get a negative variation when $\nu(t) < \theta$. For

---

[c]In the market, there exist derivative products that depend on the *realized historical* volatility of a given underlying. These kinds of products could be obviously used as a good proxy for volatility hedging, even if they are not exactly related to $\nu(t)$ that, in the case of Heston model, refers to *instantaneous* volatility.

this reason, the volatility dynamics given in Eq. (5.67) is said to be *mean-reverting*, as it continuously oscillates around its mean reversion level θ with a mean reversion speed k. For what concerns the diffusion part, parameter η drives the amplitude of the noise. It can be shown that the process $\nu(t)$ is strictly positive (as required by a volatility parameter) if the *Feller condition* $2k\theta > \eta^2$ is satisfied. In general, we observe that Heston's SDE for $\nu(t)$ is a well-known equation in QF that is usually exploited to develop interest-rate models [22–24].

In addition, we assume that there exists a correlation between the two noise terms, described by a constant parameter ρ defined as

$$dW_1(t)dW_2(t) \equiv \rho dt. \tag{5.68}$$

5.8.3. *Heston Risk Neutral Model*

A very important aspect of the Heston model is that the price of the underlying $S(t)$ is known at time t and, in particular, it does not require any forecast of the random variable $\nu(t)$. On the other hand, as the market price of risk $(\mu(t) - r(t))/\sqrt{\nu(t)}$ depends on $\nu(t)$, a unique equivalent martingale measure cannot be specified, but it will depend on the specific assumptions on the dynamics of $\nu(t)$. Unfortunately, as already mentioned above, the volatility is not a traded asset, so we cannot fix the volatility parameters by a no-arbitrage requirement as done for the underlying. As a consequence, only one equation is available to define a *set* of equivalent martingale measures that are compatible with the no-arbitrage condition.

In more detail, assuming a constant risk-free rate r, we want to exploit the two-dimensional Girsanov theorem Eq. (2.124) in order to require that the discounted underlying dynamics $\exp(-r(T-t))S(T)$ is a martingale. If, as usual, $\mathbb{P}$ is the physical measure and $\mathbb{Q}$ is the risk-neutral one, the Radon–Nikodym derivative is given by

$$\left.\frac{d\mathbb{Q}}{d\mathbb{P}}\right|_{\mathcal{F}_t}(\omega) = \exp\left[-\frac{1}{2}\int_0^t \Theta_1^2(u) + \Theta_2^2(u)du\right.$$
$$\left. - \int_0^t \Theta_1(u)dW_1(u,\omega) + \Theta_2(u)dW_2(u,\omega)\right], \tag{5.69}$$

where we defined

$$\Theta_1(t) \equiv \frac{\mu - r}{\sqrt{\nu(t)}},$$

$$\Theta_2(t) \equiv \Lambda\left(S, \nu, t\right) \equiv \frac{\lambda\sqrt{\nu(t)}}{\eta}. \tag{5.70}$$

We also observe the Brownian increment in the new measure $\mathbb{Q}$ is (Eq. (2.125))

$$dW_1^*(t, \omega) = \Theta_1(t)dt + dW_1(t, \omega),$$

$$dW_2^*(t, \omega) = \Theta_2(t)dt + dW_2(t, \omega). \tag{5.71}$$

In particular, $\Theta_1(t)$ is the equivalent BS market price of risk where we substituted the stochastic Heston volatility to the constant BS one. Moreover, in the Heston case, we have the additional parameter $\Theta_2(t)$ that we required to be equal to a generic function Λ of our model variables S, ν, t and it is a direct consequence of the market incompleteness of our model. In order to move further in our derivation, we require the Λ-function to be proportional by a constant parameter λ to the ratio between $\sqrt{\nu}$ and η.

Replacing Eq. (5.71) in Eq. (5.67), we can obtain the dynamics of $S(t)$ and $\nu(t)$ in the new measure,

$$dS(t) = rS(t)dt + \sqrt{\nu(t)}S(t)dW_1^*(t), \tag{5.72}$$

$$d\nu(t) = (k\theta - \nu(t)(k + \lambda))\,dt + \eta\sqrt{\nu(t)}dW_2^*(t). \tag{5.73}$$

In particular, we observe that defining

$$k^* \equiv k + \lambda,$$

$$\theta^* \equiv \frac{k\theta}{k^*} = \frac{k\theta}{k + \lambda}, \tag{5.74}$$

Eq. (5.73) can be restated as

$$d\nu(t) = k^*\left(\theta^* - \nu(t)\right)dt + \eta\sqrt{\nu(t)}dW_2^*(t), \tag{5.75}$$

essentially obtaining again the same expression as in Eq. (5.67), but in the new risk-neutral measure. In other words, in the Heston model, the arbitrariness due to the market incompleteness can be absorbed in model parameters and reflected in the calibration procedure. From this perspective, it does not matter which of the infinite risk-neutral measures is chosen to price instruments, as one can focus on the calibration of parameters set $(k^*, \theta^*, \eta, r, \rho)$, avoiding the need for explicitly estimating λ-parameter.

5.8.4. *Heston Pricing Formula*

Once the dynamics for the underlying in a risk-neutral measure is speci-fied, we can apply the standard pricing approach to derive a semi-analytical expression for the price of a European call option $C(S, t, \nu)$. In order to do this, we use the standard risk-neutral pricing formula and we exploit the fact that the (discounted) price of a tradable asset must be a martin-gale under the risk-neutral measure. In our framework, it can be shown that this is equivalent to requiring that the drift part of the SDE for $\exp(-r(T - t))C(S, t, \nu)$ be equal to zero (see, for example, Martingale Representation Theorem, Eq. (2.102), and recollect that the (discounted) underlying process is a martingale too). In order to do this, we first estimate the differential of the discounted price of the call,

$$\mathrm{d}(\exp(-rt)C(S, t, \nu)) = \exp(-rt)(-rC\mathrm{d}t + \mathrm{d}C), \qquad (5.76)$$

and apply the two-dimensional Itô's lemma (Eq. (2.101)), considering only the dt-term,

$$\left(-rC + \frac{\partial C}{\partial t} + rS\frac{\partial C}{\partial S} + k(\theta - \nu)\frac{\partial C}{\partial \nu} \right.$$
$$+ \frac{1}{2}S^2\nu\frac{\partial^2 C}{\partial S^2} + \frac{1}{2}\eta^2\nu\frac{\partial^2 C}{\partial \nu^2}$$
$$\left. + \eta\rho S\nu\frac{\partial^2 C}{\partial S\partial \nu} \right) \mathrm{d}t = 0. \qquad (5.77)$$

The above equation represents the Heston generalization of the Black and Scholes PDE (Eq. (5.8)) which can be solved considering the following boundary conditions:

$$C(S, t, \nu) = \max(S - K, 0),$$
$$C(0, t, \nu) = 0,$$
$$\frac{\partial C}{\partial S}(\infty, t, \nu) = 1,$$
$$rS\frac{\partial C}{\partial S}(S, t, 0) + k\theta\frac{\partial C}{\partial \nu}(S, t, 0) - rC(S, t, 0) + C(S, t, 0) = 0,$$
$$C(S, t, \infty) = S. \qquad (5.78)$$

Following Heston's original paper, we guess the solution, using a BS like formula,

$$C(S, t, \nu) = S(t)P_1 - Ke^{(-r(T-t))}P_2, \qquad (5.79)$$

where P_1 and P_2 are two unknown functions. The exact derivation of the solution is reported in Appendix A.3; here, we just report the final result:

$$P_j(x, \nu, T; \log(K)) = \frac{1}{2} + \frac{1}{\pi} \int_0^\infty \mathrm{Re}\left[\frac{e^{-i\Phi\log(K)} f_j(x, \nu, T; \Phi)}{i\Phi} \right] d\Phi, \qquad (5.80)$$

where $x \equiv \log(S)$, $j = 1, 2$, $\tau \equiv T - t$

$$f_j(x, \nu, t; \Phi) = e^{C_j(T-t,\Phi) + D_j(T-t,\Phi) + i\Phi x}, \qquad (5.81)$$

$$C_j(\tau, \Phi) = r\Phi\tau i + \frac{a}{\eta^2}\left[(b_j - \rho\eta i\Phi + d)\tau - 2\log\left(\frac{1 - ge^{d\tau}}{1 - g} \right) \right], \qquad (5.82)$$

$$D_j(\tau, \Phi) = \frac{b_j - \rho\eta i\Phi + d}{\eta^2}\left(\frac{1 - e^{d\tau}}{1 - ge^{d\tau}} \right), \qquad (5.83)$$

$$g = \frac{b_j - \rho\eta i\Phi + d}{b_j - \rho\eta i\Phi - d}, \qquad (5.84)$$

$$d = \sqrt{(\rho\eta i\Phi - b_j)^2 - \eta^2(2u_j i\Phi - \Phi^2)}, \qquad (5.85)$$

$$u_1 \equiv \frac{1}{2},$$

$$u_2 \equiv -\frac{1}{2},$$

$$a \equiv k\theta,$$

$$b_1 \equiv k - \rho\eta,$$

$$b_2 \equiv k. \qquad (5.86)$$

Equation (5.79) represents the semi-analytical solution of the Heston model for the price of a European call option. In particular, in order to estimate P_j with Eq. (5.80), a numerical integration over the real part (Re) of a complex integrand is required. We observe that the presence of the complex logarithm in Eq. (5.82) can generate discontinuities that one has to handle with suitable integration algorithms. In any case, once the complex logarithm issue is correctly managed [25], computationally fast results can be obtained from Eq. (5.79).

5.8.5.   *Heston Implied Volatility*

Heston model represents a generalization of BS model with stochastic volatility and it is usually exploited in practice in order to obtain a proper description of the volatility surface with four (calibrated) parameters k, θ, η, ρ plus the risk-free parameter r that is typically inferred from risk-free instruments and the initial ν_0. The calibration of the parameters is usually obtained requiring that the Heston prices are as close as possible to the market prices as imposed by Eq. (5.26). Once a Heston price for a call option is obtained, one can analyze this value in terms of implied volatility, inverting BS formula, Eq. (5.9). In this way, one can compare the Heston implied volatility surface with market-implied BS volatility.

In this respect, some clarifications are required, in order to avoid confusion, one should keep in mind that original BS model would imply that the whole volatility surface is perfectly *flat* as it assumes a constant volatility for each maturity and strike. In order to properly manage the volatility smile effect, one typically *modifies* the BS model, artificially introducing a *function* $\sigma(K, \tau)$ of the strike price K and the time to maturity τ. This hypothesis is clearly inconsistent with the original model assumptions and it implies, from a theoretical point of view, that the volatility of the *same* underlying depends on which point (strike price and time to maturity) we are looking at. In particular, it is not clear *which* volatility should be used for derivative contracts that, for example, lie between two strikes or depend on different times to maturity. On the other side, as the number of unknown variables, i.e. $\sigma(K, \tau)$, can be set equal to the number of equations (i.e. equal to the number of the options used for the calibration), this second implied surface assures that all market prices are perfectly matched by definition. We remark that this second implied volatility surface is not a consequence of the original BS model, as it would imply a flat surface, but it is rather how a model implied volatility should look like if the model as correct.

If we compare this implied volatility surface with the one obtained by Heston model, we can observe that, as a consequence of the stochastic volatility, the Heston implied volatility surface is not completely flat, but it naturally includes the smile effect, allowing a better fit of the market prices. In Fig. 5.6, we show an example of Heston implied volatility surface. As expected, the stochastic volatility dynamics generates a volatility smile effect that is more evident for small maturities. The amplitude of the smile is governed by the η parameter while the asymmetry mainly depends on the correlation ρ.

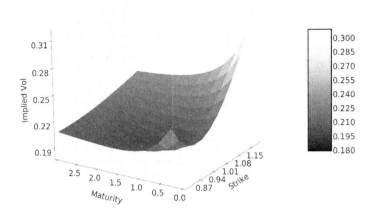

Fig. 5.6. Example of Heston implied volatility surface.

The smile effect can be analyzed considering the distribution of the underlying that for the Heston model is not Gaussian, but it exhibits fat-(exponential) tails [26] that show a better agreement with empirical data (see Section 6.6.2).

In practical applications, Heston model represents a good improvement in terms of volatility dynamics description and it is usually implemented in order to price financial instruments that are strongly affected by volatility fluctuations. In any case, in different situations, a satisfactory description of the volatility surface cannot be achieved and a generalization of the model with a larger number of parameters is required in order to obtain better fitting performances.

Chapter 6

Risk Modeling

6.1. Introduction

In this chapter, we introduce the concept of risk estimation as typically approached by banks and other financial institutions. As one can imagine, the concept of risk is quite wide and can be associated to the *perception* of something that can be dangerous; from this point of view, risk can be perceived as discretionary or judgemental. Therefore, the first two tasks that should be achieved in order to approach the risk modeling from a rational point of view are (1) to specify which kind of risks we are trying to model and (2) how we want to *measure* them. We will discuss these two issues in Sections 6.2 and 6.4. It will be shown that the probability framework is quite natural in order to deal with risky events that typically are very extreme and large in terms of magnitude. In addition, it should be noted that risk estimation is intrinsically based on the forecast of future events in probabilistic terms. We discuss this aspect in Section 6.5. This topic is a very important subject of risk management, distinguishing it from pricing which is essentially based on *risk-neutral* expectations because of no-arbitrage assumption, as shown in the Chapter 4.

In Section 6.6, we focus on market risk estimation and we present the main methodological approach used by financial institutions to estimate risk; we will extend this methodology for Counterparty Credit Risk (CCR) estimation in the next chapter (Section 7.4). Finally, in Section 6.7, it will be presented a simple discussion on how we can measure the performance of our risk models with *backtesting*.

6.2.   Risk? What is Risk?

As already stated in the Introduction, the concept of risk, and consequently of risk estimation, can be influenced by personal feelings and judgmental parameters that can be hardly included in a mathematical framework. Sometimes, the concept of risk is erroneously confused with risk aversion, that is, the reluctance of a person to accept a given level of risk. In this chapter, we will try to discern between what is a *measure* of risk and the risk aversion, and we will focus on the former. In particular, what we need to specify in order to rationally approach the problem of risk estimation are as follows:

- the scope of risk estimation, or, in other words, which kind of risks we want to measure, and
- a proper measure of risk in order to be able to compare different risks from a quantitative point of view and, as a consequence, help the management of our financial institution to take the right decision about risks hedging.

The first point is sometimes taken for granted or understood as quite a philosophical aspect; on the contrary, in most of the cases, it is more important to identify risks and be clear about what we want to measure than the risk measure itself. From this point of view, during recent financial crises, a very important role has been played by the ability of financial institutions to identify and deal with risks that had been neglected before. A typical example is credit and liquidity risk. We will further discuss about this issue in Chapter 7, where we will discuss about CCR estimation.

In this chapter, we focus on *Market risk* (MR), i.e. the risk of losses in positions arising from movements in market prices. We remark that there are not any particular reasons for this choice, nor any implied ranking of the risks; the reader should regard these two risks as good examples of application of the mathematical framework we described in the preceding chapters.

6.3.   What Kind of Model are You Looking For?

This fundamental question should be analyzed with care, as, in most of the cases, many choices about the mathematical framework used to develop the model strongly depends on the main goals we want (or we are required) to achieve. From a practical point of view, the answer to this question should help fix the constraints that our model should satisfy, and to determine which approximations we are allowed to put in place in order to simplify the problem.

In our cases, we will refer to MR (and CCR) as the risk deriving from rare events with very small probabilities, let us say less than 5%. Generally, people define these risks as *unexpected* risks in order to stress the small probability of the event. On the other hand, it would be reasonable to expect that such small probabilities would imply quite large losses that could create *serious* problems for the stability of the financial institution itself. As a consequence, the financial institutions should be ready to face these unexpected losses with a large enough amount of money in order to avoid serious consequences like default. In this financial context, it is possible to make some considerations about the expected features of the model:

- Given such extreme percentiles, questions arise on the possibility of applying the Central Limit Theorem (CLT) (Section 2.8) in order to make asymptotic assumptions about the statistics; on the contrary, Extreme Value Theory should be used. In fact, as already shown in Section 2.8, the CLT holds only in the central part of the distribution; this region can be qualitatively defined as a function of the number of the observations considered for its empirical estimation. In many practical situations, the number of such observations is not enough to assure that the CLT holds for the percentiles of interest. On the contrary, the tail behavior of the distribution (Gaussian, exponential, power-law) is the key driver of our risk measure.
- The statistical error related to our risk measure will be negatively affected by the extreme percentile requirement, so, in most of the cases, robust statistical methods should be used in order to control the statistical uncertainty and reduce the variance of the estimator.
- Given that risk estimation is meant to be used in order to face unexpected losses and avoid default, risk underestimation is considered more seriously than risk overestimation. Similarly, when it is not possible to build a model with a low level of uncertainty (typically because of data availability), a conservative risk estimation is considered the better solution. On the other hand, it has to be mentioned that risk overestimation implies that a larger amount of capital cannot be invested because it must be kept as "reserve" in order to face losses that are not likely to happen (because of the risk overestimation). In the long run, this capital freezing can have relevant impact on bank's business.

In addition, we stress that risk modeling strongly depends on the use of the obtained measure. In other situations, like capital assets allocation, where risk is considered as a driver for portfolio construction and optimization, a risk underestimation could have a strong impact on the efficiency of the portfolio itself with obvious consequences on the business. In this case, one may look at *typical* fluctuations, e.g. standard deviations, instead of rare events in order to estimate risks, following quite different modeling approaches.

These considerations are quite qualitative and should be understood as general indications on how we should set up our modeling framework and on what we could expect from our model. In the following, we will develop different modeling approaches and we will discuss the model performances in light of the above considerations.

6.4.   Risk Measures: Value-at-Risk and Expected Shortfall

Essentially, a quantitative approach is different from a qualitative one because the former is able to produce a *rational measure* of the event of interest that can be *compared* to measures of other events in order to take optimal decisions on how to manage risk. From a practical point of view, the fact that the measure is *rational* assures that it is not expert-dependent, i.e. it does not depend on people *opinion*; as a consequence, it is possible to take the optimal decision just by introducing an optimality criterion based on the obtained measures. Obviously, this theoretical framework is useful to manage risks only if

- the measure is able to correctly capture realistic *future* market dynamics on which risk can be computed, and
- the statistical uncertainty of the measure is sufficiently small in order to not negatively affect our optimality criterion.

In our case, we want to introduce a statistical function that can be used to measure the intensity of losses related to rare events, i.e. with extremely low probabilities.

In a very general framework, one could assume that there exists a probability space $(\Omega, \mathcal{F}, \mathbb{P})$ representing the future possible scenarios and $X(\omega, T)$ is a random variable on the probability space that represents the profit and loss (P&L) derived from the change of value of a given asset as a consequence of the realization of a given scenario over a specified time horizon (T). In our setup, negative realizations $x(\omega, t)$ of $X(\Omega, T)$ represent losses, while positive realizations represent profits. In the following, we will drop the dependence on time T and event ω when it is clear from the context in order to simply notation. In addition, we define the space of bounded random variables $\mathcal{S}^{\infty}$ as the space endowed with the norm:

$$||X||_{\infty} = \text{esssup}|X(\omega)|, \qquad (6.1)$$

where by esssup of a random variable Y, we denote the number $v = \inf \{z | \mathbb{P}(Y > z) = 0\}$. In other words, $\mathcal{S}^{\infty}$ is the set of random variables X for which there exists a constant $v \in \mathbb{R}$ such that $\mathbb{P}(|X| > v) = 0$.

Given this theoretical setup, we define a *risk measure* as a map $\mathcal{C}$: $\mathcal{S}^{\infty} \to \mathbb{R}$ that establishes a relation between the P&L random variable and a degree of riskiness. It would be reasonable to require that our risk measure satisfies the following conditions, known as *coherency axioms*:

- *Monotonicity* (decreasing): If a loss is smaller in absolute value than another one, the risk should be smaller:

$$X < Y \Rightarrow \mathcal{C}(X) < \mathcal{C}(Y). \qquad (6.2)$$

- *Invariance*: If one adds risk-free money in a risky portfolio, this should linearly decrease the riskiness of the portfolio itself. In our framework, risk-free money is modeled by a constant:

$$\mathcal{C}(X + a) = \mathcal{C}(X) - a \quad \forall a \in \mathbb{R}. \qquad (6.3)$$

- *Positive homogeneity*: If one doubles his exposure, his risk doubles. In general,

$$\mathcal{C}(\lambda X) = \lambda \mathcal{C}(X) \quad \forall \lambda \in \mathbb{R}^{+}. \qquad (6.4)$$

- *Sub-additivity*: It is not possible to generate extra risks (of the same type we are measuring) just by merging together different portfolios: the total risk should be less or equal to the sum of the risks of each portfolio component. Sub-additivity implies that one could obtain a *diversification benefit*, i.e. a risk reduction, in diversifying the investments in one's portfolio, which is what one would naturally expect from diversification. From a more practical point of view, this is a very important requirement in order to avoid that two almost-zero risk desks of a financial institution could generate risky investments if considered together. In formulas, sub-additivity can be expressed as

$$\mathcal{C}(X + Y) \leq \mathcal{C}(X) + \mathcal{C}(Y). \qquad (6.5)$$

On the other hand, in practical applications, quite a general class of risk measures that do not necessarily satisfy coherency axioms can be defined starting from the cumulative density functions (CDF) of the P&L F:

$$\text{Risk}_{\Xi}(F) = \int_0^1 q_u(F)\Xi(\mathrm{d}u), \qquad (6.6)$$

where

$$q_\alpha(F) \equiv \inf\{x \in \mathbb{R} | F(x) \geq \alpha\} \qquad (6.7)$$

is the *quantile* at α confidence level of the CDF F and Ξ is a (push-forward) probability measure, defined over the interval $[0,1]$. The idea underlying Eq. (6.6) is that we want to sum the losses (exploiting the quantile representation) with a given weighting scheme that is given by $\Xi(x)$; in particular, we can focus on measures defined as $\Xi(\mathrm{d}u) \equiv \varphi(u)\mathrm{d}u$. With this respect, we point out that we are not completely free to choose *any* weighting scheme, as $\Xi(x)$ must satisfy probability requirements (Section 2.2.1) and, in particular, it cannot be a decreasing function.

The two typical and mostly used choices for Ξ are as follows:

- *Value-at-Risk (VaR)*: $\Xi(\mathrm{d}x) = \delta_D(x - \alpha)\mathrm{d}x$, where $\delta_D(x)$ is the Dirac's delta function and α is a constant parameter, and
- *Expected Shortfall (ES)*: $\Xi(\mathrm{d}x) = U(x; 0, \alpha)\mathrm{d}x$, where $U(x; a, b)$ is the uniform distribution and α is a constant parameter.

Substituting these two definitions in Eq. (6.6), it is possible to obtain the standard expressions for the two risk measures, in particular for VaR:

$$\alpha \equiv \int_{-\infty}^{\text{VaR}(\alpha, F)} \mathrm{d}F(X(T)), \qquad (6.8)$$

and for ES,

$$\text{ES}(\alpha, F) = \frac{1}{\alpha} \int_0^\alpha \text{VaR}(u, F)\mathrm{d}u. \qquad (6.9)$$

From a more intuitive point of view, VaR can be equivalently defined as the maximum loss (in absolute value) within a given confidence level, expressed by $1 - \alpha$, or the minimum loss (in absolute value) we could expect with α-probability. From this definition, it should be clear that VaR implicitly gives a very sharp classification of the events we are interested in. In fact, we are very conservative in measuring risks related to events with probability ranging between α and $1 - \alpha$. In particular, we are considering the worst case, i.e. the maximum loss, as a measure of risk. On the contrary, we are completely neglecting risks deriving from events with very low probabilities, i.e. less than α, as we are including in our measure only the minimum loss related to those events. In other words, events that could generate a loss of billions could not affect our measure of risk if their probability is below the α threshold. Mathematically speaking, this behavior of the VaR measure is essentially due to the Dirac's delta function in the definition of the $\Xi(x)$ function. In fact, this measure selects only a confidence level, giving zero weight to all the quantiles different from α.

In order to avoid this problem, a natural choice would be to introduce a milder function of $\Xi(x)$ that could weight all the losses over a given confidence level. This is exactly the idea underlying the definition of the ES measure where we consider the *average* of the losses below a given confidence level; in this case, as we assumed that $\varphi(x) = U(x; 0, \alpha)$.

At this point, a very natural question would be if it is possible to find a characterization of the function $\varphi(x)$ that can assure that the coherency axioms are satisfied. The answer to this question is given by a theorem that assures that a risk measure is coherent if and only if the function $\varphi(x)$ is a density on $[0, 1]$ and decreasing with respect to x. Here, we leave aside the details of this demonstration [27, 28] and we only observe that only ES satisfies the coherency axioms, as $\varphi(x)$ is equal to $1/\alpha$ for $x \leq \alpha$ and becomes 0 for larger values of x. On the contrary, VaR is not a coherent measure and, in particular, it is not sub-additive.

From this argument, it seems natural to consider ES as a more appropriate risk measure with respect to VaR. Actually, the situation is more complex and the problem is quite debated by academics and practitioners. In particular, a crucial point is related to the robustness of the statistical estimation of these measures that obviously depends on the model considered to estimate them. Qualitatively, one should firstly observe that, for a given confidence level α, the ES measure implies a *larger* risk than VaR estimation. And, by definition, the first measure refers to an average of the losses below α, while the second measure represents the minimum of the same losses sample. So, in order to compare the robustness of the two measures, one needs to define a proper percentile level, different for each measure in order to consider the same level of risk. As a second step, one should define a robustness measure that could take into account the statistical uncertainty of our estimation. Intuitively, one could expect that for the VaR, the statistical uncertainty depends on the chosen percentile level, i.e. the lower the α the higher the uncertainty, as typically information decreases as the event becomes rarer. With respect to ES, the situation is a little bit more complicated as we are considering two competitive effects. From one side, we are taking the *average* of a given sample losses, so our estimation should be more stable because of the CLT. On the other side, in our sample, we have losses related with extremely low percentile levels that are typically affected by high uncertainty. So, the robustness of the ES estimation will be affected by these two effects in a non-trivial way that ultimately depends on the loss distribution.

From these simple deductions, it should be clear that it is quite difficult to obtain a definitive answer about the best measure to be used, as the final result could be affected by model-dependent factors. In this respect, very interesting results can be found in Ref. [29] where, for historical estimation, it is shown that VaR is robust in contrast with ES. From a practical perspective, the best choice in our opinion is to provide the management both alternatives in order to obtain a more complete description of risk. In fact, in most of the cases, the computation effort required for risk estimation is mainly related to the construction of the P&L distribution, while risk measures can be computed quickly.

In the following, we will focus mainly on VaR estimation, even if most of the considerations apply to ES as well.

6.4.1. *Some Caveats on VaR Estimation*

In this section, we outline some caveats about VaR estimation that we think can be useful in practical situations. Specifically, most of the problems we are going to discuss do not have a simple solution or their solution does not exist at all; they remain an open issue in risk modeling. These situations are quite frequent in any modeling activity where simplifications must be introduced. As a consequence, a *model error* will affect measures. In these cases, it is worthwhile to study these errors in order to monitor them, having a clear understanding of how they are generated.

In this section, we analyze in detail some caveat about VaR estimation, assuming that the P&L distribution of a financial instrument (or a portfolio of financial instruments) was already obtained by a given methodology. Furthermore, as it typically happens in practical situations, we will assume that the P&L distribution is not theoretically known, but it can be empirically deduced considering a sample of n P&L observations. In other words, we are focusing on the percentile estimation of a given empirical distribution, given the percentile level α and the set of P&L observations $x_1, \ldots, x_n$. In particular, in the following sections, we focus our attention on two very important aspects of risk management:

- statistical percentile estimation, and
- statistical uncertainty of VaR.

6.4.1.1. *Percentile Estimation*

If the number of observations is very large, one can make use of the CLT and standard percentile estimator to calculate VaR. In particular, one can define the *empirical* percentile estimator at α-level as

$$pct(\alpha) = x_{\lfloor n\alpha \rfloor}, \tag{6.10}$$

where x_k is the kth least element of the set $\{x_i\}_{i \leq n}$ and $\lfloor a \rfloor$ denotes the integer part of a. With this definition, we are implicitly assuming that the probability of getting an observation lower than x_1 is $1/n$ and the probability of getting an observation larger than x_n is zero. On the other hand, one could define the estimator of the percentile so that the probability of getting an observation lower than the minimum is zero and the probability of getting an observation *greater* than x_n is $1/n$:

$$pct^+(\alpha) = x_{\lfloor n\alpha+1 \rfloor}. \tag{6.11}$$

This arbitrariness introduces an uncertainty in the percentile estimation. From a risk management point of view, the definition given by Eq. (6.10) is more conservative, as by definition $x_{\lfloor n\alpha \rfloor} < x_{\lfloor n\alpha+1 \rfloor}$, so it should be preferred in the case of uncertainty.

In any case, it is possible to prove that, *when the CLT holds*, the difference between the empirical estimator (Eq. (6.10)) of the percentile and its theoretical value converges to a normal distribution of zero mean and variance given by (see Ref. [30] for further details)

$$\mathbb{E}[pct(\alpha)] = u(\alpha),$$

$$\text{var}(pct(\alpha) - u(\alpha)) = \frac{\alpha(1-\alpha)}{np^2(x)}, \tag{6.12}$$

where $p(x)$ is the theoretical probability density function (PDF) of the variable x and $u(\alpha) \equiv F^{-1}(x; \alpha)$. As a consequence, we are allowed to use the standard percentile estimator given by Eq. (6.10) whenever we can prove that the CLT holds.

From another point of view, one could wonder *how* the convergence implied by CLT is reached as we move from an external region to the central part of the distribution, where we know that the CLT holds; in this respect, the notion of bias estimator is of crucial importance (see Sections 2.6 and 2.8). From the previous considerations on the percentile estimators (Eqs. (6.10) and (6.11)), one could expect that the estimator defined in Eq. (6.10) is biased, thus underestimating the true percentile (obviously the contrary should be valid for Eq. (6.11)) even if this guess could seem in contrast with Eq. (6.12). Indeed, one could check this hypothesis by a numerical simulation repeating a large number of times the empirical estimation of the percentile on different (Gaussian) random samples of length n. Knowing the theoretical distribution, one can compare the theoretical value of the percentile $u(\alpha)$ with the sample mean of the historical estimations of its $pct(\alpha), pct^+(\alpha)$.

In Fig. 6.1, we show the comparison between theoretical and empirical percentile for a Gaussian random sample as a function of n, considering the two definitions of the percentile estimator. From the figure, it is clear that the two percentile estimators are biased, even if this result seems to be in contrast with the previously described Eq. (6.12). In this respect, the crucial aspect that should be kept in mind is that, when we apply the CLT, we are taking the limit for $n \to +\infty$ and for a *fixed* percentile level. In other words, when we increase n, we are increasing the range of probabilities spanned by our sample that roughly goes from $1/n$ to $1 - 1/n$; as a consequence, our fixed percentile level α moves through the middle of this range. This fact is

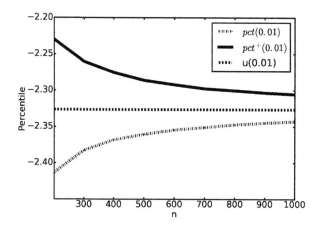

Fig. 6.1. We show the percentile bias of the estimators of Eqs. (6.10) and (6.11) for a confidence level $\alpha = 0.01$ as a function of the number of observations n. As evident from the figure, as n increases, the two estimators converge to the theoretical value $u(0.01)$.

in good agreement with the reduction of the distance between the estimator and the theoretical value shown in Fig. 6.1. The additional information that is represented in the figure but is not given by the CLT theorem is *how* fast the convergence is reached by the estimator, as already stated above.

In particular, one could be interested in what is a suitable value of α so that the difference between the measure and the theoretical value is small.

Indeed, it is possible to analytically estimate the critical value of α, so that the convergence is assured. To do this, one should consider, for example, the demonstration of the theorem shown in Ref. [30] and cited above. Here, we do not report the whole demonstration as it is quite technical. We only observe that a crucial passage is related to the application of the CLT to the quantity $\sqrt{n}(C_n/n - \alpha)$ to obtain convergence to $N(0, \alpha(1 - \alpha))$ for $n \to +\infty$, where C_n is the number of the observations in our sample less or equal to $u(\alpha)$. It can be shown that the variable C_n is distributed as a binomial PDF with mean $n\alpha$ and variance $n\alpha(1 - \alpha)$. In other words, our original problem of determining how the percentile estimator converges to its theoretical value leads to defining the condition such that the C_n random variable can be well approximated by the normal distribution $N(n\alpha, \sqrt{n\alpha(1 - \alpha)})$. To this aim, we can try to extend a very interesting result in Ref. [31] where it was shown that is possible to estimate the critical number of observations to apply the CLT. In particular, following the lines of the demonstration we define $X = \sum_{i=1}^{n} x_i$ as the sum of n i.i.d. random variables of mean μ, variance σ^2, and skewness λ_3 and

$$y \equiv \frac{X - n\mu}{\sigma\sqrt{n}}, \tag{6.13}$$

and we estimate the error ΔE approximating the distribution of y with a normal distribution as

$$\Delta E = F_G(y) - F(y) = \frac{\exp(-y^2/2)}{\sqrt{2\pi}} \left(\frac{Q_1(y)}{n^{1/2}} + \cdots + \frac{Q_k(y)}{n^{k/2}} + \cdots \right), \tag{6.14}$$

where $F_G(y)$ is the Gaussian CDF and $Q_k(y)$ are polynomial functions which can be expressed in terms of the normalized cumulants λ_k defined as

$$\lambda_n \equiv \frac{k_n}{\sigma^n}, \tag{6.15}$$

where k_n is the n-cumulant as described in Section 2.5. We remark that the normalized cumulant appearing in the expansion refers to the original variable x_i.

The first leading term of the expansion in Eq. (6.14) is

$$Q_1(y) = \frac{1}{6}\lambda_3(y^2 - 1). \tag{6.16}$$

So, from this simple description, to reduce the error of Eq. (6.14), allowing the application of the CLT, one should require that $\sqrt{n} \gg Q_1(y)$ or $n \gg \lambda_3^2$. On the same lines, if $\lambda_3 = 0$, the second term of the expansion would lead to the requirement $n \gg \lambda_4$. These simple results show that λ_3 and λ_4 are two indicators of how much a generic distribution is not Gaussian. In particular, to reduce the error of Eq. (6.14) making possible the application of the CLT, one should require that $n \gg \lambda_3^2$.

In our case, we need to estimate λ_3 for independent and identically distributed random variables that summed n-times generate C_n. In particular, from C_n definition, we consider n Bernoulli distributed (Section 2.7.2) random variables x_i, $i = 1, \ldots, n$ each with α probability to be extracted, i.e. $x \in Ber(x, \alpha)$ and we focus on $C_n = \sum_{i=1}^{n}$. It can be shown that for a Bernoulli distribution,

$$\lambda_3^2 = \frac{(1 - 2\alpha)^2}{\alpha(1 - \alpha)}, \tag{6.17}$$

therefore, the condition to obtain the convergence of the percentile estimation is given by

$$\frac{(1 - 2\alpha)^2}{\alpha(1 - \alpha)} \ll n, \tag{6.18}$$

or

$$\lim_{n \to +\infty} \frac{1}{n} \frac{(1 - 2\alpha)^2}{\alpha(1 - \alpha)} = 0. \tag{6.19}$$

Using Eq. (6.18), one can estimate a suitable value of n to obtain the convergence of the estimator. For example, the convergence is not reached if $\alpha = 1/n$, i.e. if we are considering a percentile level that refers to the worst loss in our sample. In that case, the CLT and, in particular, Eq. (6.12) do not hold, so the percentile estimator is biased and a faithful risk estimation cannot be obtained. On the contrary, for a *fixed* value of α, the convergence is always reached as n increases. In general, in common Risk Management applications, the value of n is often influenced by the low availability of market data and by the general requirement of using data representative of the current market framework; as a consequence, the condition of Eq. (6.18) cannot be met in every situation. For this reason, one could be interested in slightly modifying the historical percentile definition to obtain an unbiased estimator for a fixed confidence level α and a low number of observations n.

From this point of view, some insight can be obtained by the application of the Extreme Value Theory (EVT) in Refs. [32, 33]. This very fascinating theory analyzes the problem of characterizing the distribution of extremely rare events and it can be understood as the theory that should be used when the CLT does not hold, i.e. in the tails of the distributions. In a nutshell, the main EVT result is that rare events can be classified in three classes according to the tail behavior of the distribution they belong to. In particular, there exists a one-to-one relation between the tail behavior of the random variable and the distribution of its rare events. In Table 6.1, we summarize this result. In particular, we generically refer to distribution *family* in order to stress that a detailed characterization of the distribution is not needed in EVT and only the tail characterization is enough to describe the low probability events.

In Table 6.1, we report how the minimum of a historical series of a large number of observations is distributed, when the tail behavior of the original random variable distribution is known. In our experience, it is quite easy to make confusion between the original distribution of the random variable and the distribution of the rare events. To clarify the situation, we suggest the following example.

First, we consider n realizations of *exponentially* distributed random variables. We ask ourselves: *how should the empirical distribution look like?* In this case, the answer is quite trivial, i.e. exponential. Now, consider m samples of n realizations each and take for each sample the minimum. In this way, you have m realizations of the minimum of a sample of n observations.

Table 6.1   Distribution of the minimum of a large number of observations with respect to the distribution family

Distribution Family	Example	Prob. Smallest Value (Asymptotic)
Exponential	$F(x) = \exp(\gamma x)$ $\gamma > 0, x \leq 0$	Gumbel $F(z) = 1 - \exp\left[-\exp-\gamma(z-u)\right]$
Pareto (power-law)	$F(x) = (\omega - x)^{-\gamma}$ $x < \omega - 1, \gamma \geq 1; \omega \leq 0$	Frechet $F(z) = \exp\left[-\left(\frac{z-\omega}{u-\omega}\right)^{-\gamma}\right]$
Limited	$F(x) = (\omega + x)^{\gamma}$ $-\omega \leq x < 1 - \omega; \gamma > 1; \omega > 0$	Weibull $F(z) = \exp\left[-\left(\frac{\omega-z}{\omega-u}\right)^{\gamma}\right]$

EVT answers the question: *how are the minima distributed?* From Table 6.1, we can see that the answer is a Gumbel distribution, assuming that n is large enough. In Table 6.1, the γ parameter does not depend on n as we considered a specific distribution; in a more general framework, γ may depend upon n.

For what concerns the exponential distribution, a very interesting result is that the probability of obtaining a value smaller than the minimum of the sample is always given by $1 - 1/e \sim 63\%$, i.e. quite a high probability. This fact stresses the difficulties, from a statistical point of view, of obtaining a reliable estimation of the minimum loss given a historical series.

In general, it is possible to extend this EVT result for extreme values (not necessarily the minimum or the maximum) of a distribution as it is required for VaR estimation. In the following, we will focus on exponential and power-law tails as they are the most representative of P&L distributions. Using EVT, it is possible to show that for exponential PDFs, the following relation holds:

$$\mathbb{E}[pct(\alpha)] = u(\alpha) + \frac{y_m}{\beta}, \tag{6.20}$$

where

$$m \equiv \lfloor n\alpha \rfloor, \tag{6.21}$$

$$y_m \equiv -\ln(m) + \sum_{v=1}^{m-1} \frac{1}{v} - \gamma, \tag{6.22}$$

$$\beta(\alpha) = \frac{p(u(\alpha))}{F(u(\alpha))}, \tag{6.23}$$

and γ is the Euler–Mascheroni constant. In particular, β is the so-called *extremal intensity function* or hazard rate. For example, for Gaussian and Exponential ($p_{\exp}(x) = \exp(\gamma x), x \le 0$) PDFs, one has

$$\beta_{\text{Gauss}}(\alpha) = \frac{p(x(\alpha))}{1 - F(x(\alpha))} \sim \frac{\dfrac{e^{-\frac{x^2(\alpha)}{2\sigma^2}}}{\sqrt{2\pi\sigma^2}}}{\dfrac{e^{-\frac{x^2(\alpha)}{2\sigma^2}}}{\sqrt{2\pi\sigma^2}} \sigma/x(\alpha)} = \frac{x(\alpha)}{\sigma^2}, \tag{6.24}$$

$$\beta_{\exp}(\alpha) = \gamma. \tag{6.25}$$

We remark that even if $1/\beta_{\text{Gauss}}(\alpha)$ is an increasing function of α, the ratio $y_m/\beta(\alpha)$ is still a decreasing function, so a larger percentile level would imply a larger convergence of the empirical percentile estimator as one would expect.

Equation (6.20) has a very important meaning as it describes the relation between the expected value of the percentile estimator (Eq. (6.10)) and the theoretical value of the percentile $u(\alpha)$ for confidence level α. In particular, as y_m is a negative function, the expectation of the percentile estimator is always lower than the theoretical percentile value. This fact confirms theoretically what we had already observed when we defined the percentile estimator, i.e. the fact that Eq. (6.10) leads to a conservative risk estimation. In addition, knowing the P&L distribution family and the intensity parameter, one could correct the bias in the estimator by the quantity y_m/β in order to recover the theoretical percentile value.

In general, one can also obtain the complete moment generating function (Eq. (2.38)) of $pct(\alpha)$ for an exponential distribution:

$$G_{pct(\alpha)}(t) = e^{u(\alpha)t}\frac{m^{-t/\beta(\alpha)}\Gamma(m + t/\beta(\alpha))}{(m-1)!}. \tag{6.26}$$

With this relation, it is possible to estimate for this distribution family all the moments related to the mth extreme value. In particular, Eq. (6.26) can be used to estimate the statistical uncertainty of an empirical VaR estimation.

Similar relations hold for power-law decaying PDFs. In this case, the relation between the expectation of the percentile estimator $pct^*(\alpha)$, the theoretical percentile $u^*(\alpha)$, and the extremal intensity function β^* can be obtained applying the following transformation to the previous relations:

$$u = \ln(u^*), \tag{6.27}$$

$$pct(\alpha) = \ln(pct^*(\alpha)), \tag{6.28}$$

$$\beta = \beta^*, \tag{6.29}$$

where β^* is defined as

$$\beta^* = \lim_{z \to +\infty} z\frac{f(z)}{1 - F(z)}. \tag{6.30}$$

In particular, for power-law decaying PDFs, one has

$$f(x) = \frac{\mu A^\mu}{|x|^{1+\mu}} \qquad \text{for} \quad x \to \pm\infty, \tag{6.31}$$

$$F(x) = 1 - \frac{\mu A^\mu}{x^\mu} \qquad \text{for} \quad x \to +\infty, \tag{6.32}$$

where A gives the order of magnitude of the tail. From the previous equations, we obtain

$$\beta^* = \mu. \tag{6.33}$$

Note that in this case, as for Eq. (6.25), the extremal intensity function does not depend on $u^*(\alpha)$ for sufficiently large u^*. The moments of $pct^*(\alpha)$ can be estimated with the following relation:

$$\mathbb{E}[pct(\alpha)^l] = G_{\log(pct(\alpha))}(l). \tag{6.34}$$

In particular, using Eq. (6.26), we have

$$\mathbb{E}[pct^*(\alpha)] = u^*(\alpha) \frac{m^{1/\beta^*(\alpha)} \Gamma(m - 1/\beta^*(\alpha))}{(m-1)!}. \tag{6.35}$$

From Eq. (6.35) (analogous to the exponential family case), one can correct the bias in the empirical percentile estimator, assuming that it is possible to estimate $\beta(\alpha)$. In Fig. 6.2, we show the comparison between the expected value of the empirical percentile estimator (Eq. (6.10)) and its version with the bias correction (inverting Eq. (6.35)) for a sample of $n = 500$ observations coming from a Student's t-distribution. The convergence is improved using EVT estimator assuming that the number of degrees of

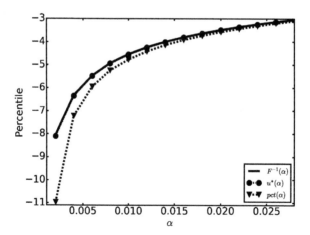

Fig. 6.2.   We show the comparison between the expected value of the empirical percentile estimator (Eq. (6.10)) and its version with the bias correction (inverting Eq. (6.35)) for a sample of $n = 500$ realizations coming from a Student's t-distribution. As the confidence level α increases for a fixed number of observation, the convergence increases. For small percentile confidence levels, the standard estimator $pct(\alpha)$ is bias, while the EVT estimator ($u^*(\alpha)$) converges to the theoretical value $F^{-1}(\alpha)$.

freedom of the distribution is known. Similar results also hold for Gaussian and Exponential distributions.

Unfortunately, this approach can only be applied if β parameter, which describes the tail behavior of the theoretical PDF, is known. In the absence of a theory that could suggest what the correct tail behavior is, one could try to estimate it, considering the available observations. In any case, the characterization of tail behavior is not an easy task, as will be discussed in Section 6.6.2. As an alternative, as already mentioned above, one could consider conservative choices, as the one given by Eq. (6.10), keeping in mind that when the CLT does not apply, differences from the percentile theoretical value could be non-negligible.

6.4.1.2. *VaR Statistical Error*

Another caveat to keep in mind in the case of percentile estimation is the statistical uncertainty of this estimator, especially when the required percentile level is comparable to the minimum probability spanned by the sample, i.e. $1/n$, where n is the number of the observations.

As already discussed, when the CLT holds, we already have an estimation of the variance of our estimator, given by Eq. (6.12). With respect to this equation, we point out the following considerations:

- The PDF $p(x)$ in the denominator refers to the *theoretical* distribution of random variable x. In practical situations, the theoretical distribution is typically unknown (otherwise, there would not be reasons to estimate the percentile empirically). As a consequence, the variance of our estimator will be subject to some approximation in the empirical estimation of $p(x)$. On the same lines, the value of x must be the theoretical value $u(\alpha)$ of the percentile.
- Since the CDF the integral of the PDF, it can be shown (for example, by applying De l'Hopital theorem) that the variance of the estimator increases as x becomes large in absolute value (provided that $\frac{dp(x)}{dx}$ goes to 0). This behavior is quite reasonable, as, for bell shaped distributions, the larger (in absolute terms) the x value, the rarer the event and the lower the information about the probability of this event. As a consequence, the uncertainty of the estimator is high.
- As one could naturally expect, as the sample size n increases, the uncertainty about the percentile estimation *for a fixed probability level* α decreases.

Unfortunately, Eq. (6.12) is valid only when the CLT holds. In analogy with the previous section, one can use EVT to obtain a reliable estimation of the percentile uncertainty, assuming that the P&L distribution family is known. In this case, the second moment of $pct(\alpha)$ for exponential and power-law distributions can be estimated using Eqs. (6.26) and (6.34) and it can be used to evaluate the uncertainty of the percentile estimator,

$$\mathbb{E}\left[(pct(\alpha))^2\right] = \frac{1}{\beta(\alpha)} \sum_{k=m}^{\infty} \frac{1}{k^2}, \tag{6.36}$$

$$\mathbb{E}\left[(pct^*(\alpha))^2\right] = \frac{(u^*(\alpha))^2 m^{\frac{2}{\beta^*}}}{(m-1)!}, \tag{6.37}$$

$$\left(\Gamma(m - \frac{2}{\beta^*}) - \frac{\Gamma^2(m - \frac{1}{\beta^*})}{(m-1)!}\right), \tag{6.38}$$

where $\Gamma(x)$ is the gamma function.

As an example, we simulated a sample of $n = 500$ P&L observations generating random variables from a Student's distribution with $\nu = 3$ degrees of freedom, and we calculated VaR considering the estimators given by Eq. (6.10) and inverting Eq. (6.35) for different percentile levels α. Then, we repeated this experiment a large number of times, and compared the numerical variance of the percentile measure with the theoretical value that can be computed with Eqs. (6.12) and (6.36). In Fig. 6.3, we report the results of our analysis. As one could expect, as the percentile level α increases, i.e. m increases for a fixed value of n, the numerical variance of the estimator converges to the value that can be obtained assuming that the CLT holds, Eq. (6.12). On the contrary, for very extreme percentiles, the CLT values are not in agreement with the numerical results, and EVT applies.

From a practical perspective, this simple example shows that when an empirical percentile estimator is considered for VaR estimation purposes, its statistical uncertainty should be taken into consideration in choosing the number of observations and the desired percentile level.

6.5. Real-World Measure

In Chapter 4, we demonstrated that the fundamental probability measure that should be used for pricing purposes is the risk-neutral one. In a nutshell, the reason for this choice is that we do not need to forecast future movements of the market, but we are trying the obtain an

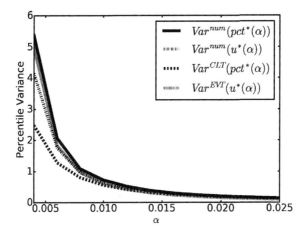

Fig. 6.3. We show the variance of the two percentile estimators (Eqs. (6.10) and (6.35)) obtained by numerical simulation (subscript *num*), the CLT and the EVT approximations as a function of the confidence level α. It can be observed that for small percentile levels, the variance of the two estimators is much larger than the one estimated by CLT.

appropriate replica of the payoff of the considered instrument, hedging risk and obtaining a risk-free performance. In other words, pricing is not a matter of forecasting prices but of comparing performances.

On the other side, risk modeling is completely different as we are interested in estimating typical *future* risky events in order to quantify possible losses. In other words, risk modeling aims to answer the question: *What happens if we do not hedge the risks?* In this case, we are not allowed to use the risk-neutral measure as we are not replicating a product. As a consequence, in order to estimate risks, we have to rely on the so-called real-world (or physical) measure. This aspect is very important, and it determines the fundamental difference between pricing and risk modeling. In particular, one cannot switch from the drift term to the risk-free term.

As a consequence of the different modeling aims, a key difference between risk modeling and pricing is that calibration is based on historical data and not on the current market quotes.

For example, in the risk world, the volatility of an underlying is typically estimated considering historical series analysis. On the contrary, from a pricing perspective, implied volatility is used (see Section 5.4). This fact has different implications in the two worlds. From a pricing perspective, the main requirement is to match the current market prices like call and put options for the Black–Scholes (BS) model. This is a very crucial point as otherwise the price of your portfolio could imply a gain or a loss that cannot be realized in the market. As a consequence, volatility parameter is chosen in order to achieve this goal, considering the current price of call and put options. From the risk modeling perspective, the main goal is to forecast the future risk-factor fluctuations, especially rare events that could generate large losses. Therefore, time series analysis is typically preferred in order to identify stable statistical behaviors that can be assumed to be representative of the future. This is in contrast to the implied volatility estimation that is based only on the current market price.[a] This example shows one of the key differences in modeling techniques used in a risk-neutral and real-world context. In the following, we discuss several approaches used for MR.

6.6. Market Risk Estimation

In the preceding sections, we discussed the risk estimation from a general perspective and we introduced the typical measures used for risk estimation. In this section, we focus our attention on the *MR*, i.e. *the risk of losses in positions arising from the movements of MR factors.*

From a general perspective, there are no particular reasons that justify the focus on this risk with respect to other sources of risk, nor is risk ranking somehow implied by this choice. In fact, one could naturally expect that the relevance of a risk type strongly depends on the core business of the bank or financial institution considered. On the other side, from a purely modeling perspective, the huge availability of market data makes MR modeling a very attractive field where people with a quantitative background can be involved. In fact, in our view, sophisticated and mathematically interesting models can be developed only when data availability assures the opportunity to test the hypotheses introduced from an empirical point of view. On the contrary, when data are not available, the easier and most conservative choice should be made. This typically makes the model less interesting from a mathematical perspective, but more robust and able to manage the high uncertainty implied by the poor dataset.

---

[a]In this respect, one could try to mix the two techniques somehow, i.e. the current price matching and the time series analysis, in order to derive an appropriate indicator of future risk-factor fluctuations.

In the following sections, several approaches will be discussed for MR estimation. Here, we observe that from the definition of MR two key elements can be highlighted:

- movements of MR factors, and
- losses, i.e. negative variations of the price of the considered financial instruments.

For what concerns the first point, we need to specify for each product which MR factors[b] $S(t)$ could affect its price. By MR factors, we generally mean the factors that could represent a risk and that can be objectively deduced from market information. For example, for a call option, a very important MR factor is the price of the underlying as its variation could generate a variation in the call price. As a consequence, a very important activity consists in the mapping of each financial instrument (in the considered portfolio) in an appropriate set of risk factors. Needless to say, an incorrect mapping would lead to a misrepresentation of MR. Once risk factors are identified, a definition of market *movements* $\Delta S(t)$ is required. In general, we can assume that we have a dataset that represents a path of the risk-factor dynamics made by n observations $\{S(t_1), \ldots, S(t_n)\}$, and we want to define a suitable increment in order to model the risk-factor movements. At least, three choices are available:

$$\Delta S(t) \equiv S(t + \Delta t) - S(t),$$

$$\Delta S(t) \equiv \frac{S(t + \Delta t) - S(t)}{S(t)},$$

$$\Delta S(t) \equiv \log \frac{S(t + \Delta t)}{S(t)}, \qquad (6.39)$$

where Δt is a suitable time increment. In general, we will refer to these three alternatives as *absolute, relative and log*-variations, respectively. Obtaining a general result about what is the best choice for the risk-factor movements is not an easy task and, in general, depends on the considered risk factor. As a general rule, one should choose a function $S(t)$ that can transform it in identically distributed random variables. In this way, one can assume that each random variable of the (transformed) time series

---

[b]In this context, we used for the risk factor the same notation we considered for the underlying in the pricing framework. Actually, in the risk modeling context, the concept of risk factor can assume quite a wider meaning, not necessarily referring to the underlying price.

$\{\Delta S(t_1), \ldots, \Delta S(t_{n-1})\}$ is actually a realization of the same distribution. As a consequence, this new dataset can be used for statistical purposes, for example, to infer the moments of a theoretical distribution or to bootstrap a new distribution starting from the original dataset. In any case, one can generate sequences of risk-factor movements that can be *applied*[c] to the current value of the risk factor in order to obtain a distribution of possible future scenarios. For example, let us assume that our risk factor follows lognormal (BS like) dynamics as described by Eq. (5.1)

$$dS(t) = \mu S(t)dt + \sigma S(t)dW(t). \tag{6.40}$$

Assuming this dynamics, a bad choice would be to consider absolute increments as evident from the discretized version of the equation,

$$\Delta S(t) = \mu S(t)\Delta t + \sigma S(t)\Delta W(t), \tag{6.41}$$

that the absolute increment *depends* on the current state of the variable $S(t)$. In particular, the higher the value of the risk factor $S(t)$, the higher the absolute increment that we should expect given a fixed time interval Δt. As a consequence, in this case, absolute increments *cannot* be assumed to be identically distributed, and it would be difficult to estimate μ, σ using standard statistical estimators. On the contrary, considering log-variations, by application of Itô's lemma, one can obtain the dynamics given by Eq. (4.2) that we consider here in its discretized version,

$$\Delta z(t) = \left(\mu - \frac{\sigma^2}{2}\right)\Delta t + \sigma \Delta W(t). \tag{6.42}$$

From this equation, it is evident that all the lognormal increments, that can be obtained from our dataset, belong to the same distribution for a fixed time increment, and they can generate new risk-factor scenarios. For example, one could use sample mean and standard deviation estimators to obtain an estimation of the parameters of the model, generate new risk-factor variations distributed as a Gaussian with these parameters and finally apply them to the current risk-factor value in order to obtain the future distribution of the risk factor.

From this simple example, it should be clear that the modeling choices about market increments could strongly affect our final risk estimation.

---

[c]The application of the risk-factor movements to a starting risk-factor value depends on the definition of the movement itself and on the modeling choices. For example, if we are considering absolute variations, we could apply a risk-factor movement just by adding it algebraically.

Another peculiar fact related to absolute variations is that applying them (by algebraic sum) to a positive risk factor, we could end up in a negative value. For this reason, the absolute variation should not be used for risk-factors which are positive by definition, e.g. underlying prices.

Once risk-factor fluctuations are properly defined, one needs to estimate how much they can affect the value of the financial instruments of interest. This aspect is obviously related to the second element of MR modeling, i.e. the estimation of possible losses. At this point, the role of the pricing theory in the risk management context should be clear: once different risk-factor scenarios are generated, one needs to evaluate for each configuration the pricing function related to the financial instrument of interest; in this way, it is possible to move from the distribution of future risk factors to the distribution of future prices of the financial instrument, from which a risk measure can be deduced. In this framework, the role of the real-world and risk-neutral measure is well defined and interrelated: the former is used for generating risk-factor scenarios, the latter is used for estimating the price for each scenario configuration.

From a computational perspective, it should be clear at this point that computational effort required for risk management purposes is generally higher than for pricing, and a dedicated computation environment may be required, typically leveraging also on parallel-computing mechanisms. The reason is that one needs to estimate prices for many different scenarios to build a price distribution, and this obviously increases computational complexity.

In some cases, some approximation in the pricing model may be required in order to speed up the computation. In this respect, a typical example is the sensitivity approach based on Taylor's expansion and using the Greeks for the estimation of price fluctuations. In particular, following the same approach already discussed for hedging in Section 5.7 and considering Eq. (5.63), it is possible to obtain the distribution of price variations given the sensitivities at each risk factor (estimated only once) and the distribution of risk factor variations without the need to apply the pricing function to each scenario. This approach is computationally less demanding, but it represents a good approximation as long as risk-factor variations are small. This fact could represent an issue as in most of the cases, a large variation of the risk factor would imply a large variation in the price, i.e. large risk. As a consequence, it could lead to a misspecification of the risk estimation.

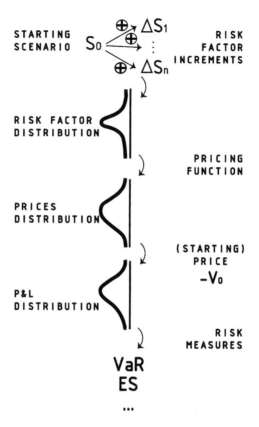

Fig. 6.4. A representation of the MR estimation algorithm.

In conclusion, we can summarize our risk modeling approach as composed by the following steps (see Fig. 6.4):

- identify the main risk factors for each financial instrument,
- define a suitable random variable (risk-factors variations) in order to deal with identically distributed realization,
- define an algorithm to generate future risk-factor variations, and apply them to the current risk-factor value to obtain a future distribution of risk-factor values,
- using appropriate pricing functions (or suitable approximations), transform risk-factor distribution into future price distribution and take the difference with the current instrument price to obtain the P&L distribution,

- estimate one or more risk measures using the obtained P&L distribution,
- repeat this algorithm for each instrument in your portfolio.

In this section, we discussed in detail the estimation of a risk measure *given* a P&L distribution. In the following sections, we will describe several typical approaches for the construction of the risk-factor distribution, assuming that for each instrument in our portfolio, an appropriate pricing function is given.

6.6.1. *Historical Method*

In this section, we are going to discuss one of the most used methods for the construction of future risk-factor distribution, i.e. the historical approach. It aims to exploit the information available in the risk-factor time series, e.g. the past prices of option underlying instruments. A typical example is to consider the price time series of an equity, e.g. IBM, and take into consideration the relative variation of prices as defined in the previous section, *assuming that each variation* $\Delta S(t_i)$ *is an independent and identically distributed random variable.* Without any additional assumption, one can obtain the empirical distribution of risk-factor variations as already described in Section 2.7.4, and, applying each variation to the current value of the risk factor, generate a new distribution that we assume represents the future distribution of the risk-factor values. Given the initial set of prices $\{S(t_1), \ldots, S(t_n)\}$, where n is the total number of realizations, one can express the risk-factor variation in relative terms, obtaining the set of relative variations $\{\Delta S(t_1), \ldots, \Delta S(t_{n-1})\}$. Applying each variation to the last available price (we assume that t_n is today), we finally obtain the set,

$$\{S_f^1 = S(t_n)(1 + \Delta S(t_1)), \ldots, S_f^{n-1} = S(t_n)(1 + \Delta S(t_{n-1}))\}, \quad (6.43)$$

where S_f^i is the ith realization of the risk factor.

The main advantage of this method is that no explicit assumption on the theoretical distribution of risk factors is required, meaning that it is not necessary to make explicit the relation between the empirical data and theoretical distributions. In particular, all the "shocks" applied to the last available price can be justified in historical perspective, as they actually happened in the past. In this respect, another big advantage of the historical approach is that if the simulation of many *correlated* risk factors were required, the explicit estimation of the correlation matrix would not be needed, since correlation is automatically reflected in their historical series. This aspect is of fundamental importance for at least two reasons: (1) the estimation of the correlation is not an easy task and it typically ends up with quite a large statistical uncertainty that could negatively affect the risk measure, and (2) as the risk-factor number is typically large for portfolios, the storage of the correlation matrix would require a large amount of computer memory with negative effects on the performances.

On the other hand, the main hypothesis underlying this approach is that the random variables representing risk-factor variations are actually independent and identically distributed in order to justify the general idea "picking up" something from the past and projecting it into the future. In other words, the historical method implicitly assumes that in some sense,[d] the history is cyclic and will repeat itself in the future. This could be analyzed on a more philosophical level, and one could ask if the analysis of historical series is well suited for disciplines, like Quantitative Finance (QF), that are strongly related with human behavior that can be, for its nature, completely incoherent through the time. Here, we leave aside this theoretical aspect, and we focus on a more practical approach, observing that the accuracy of the model is strongly related to the efficiency in the construction of i.i.d. historical random variations. For example, we have already shown in Fig. 5.3 the relative variations of IBM prices. After a careful observation of the graph, it is evident that the intensity, i.e. the absolute value, of the variations is not constant in time, but it seems to be organized in clusters alternating high and low levels of fluctuations. This effect is known as *volatility clustering*, and it is recurrent in historical series. This observation is an empirical evidence that volatility is not constant in time and, in general, it exhibits *heteroskedastic* effect. In addition, some fluctuations are much larger (or much smaller) than their typical magnitude in a given time period. These large fluctuations are typically caused by very important market news that affected the considered asset. These kinds of fluctuations can be described as spikes if the price comes back to the original value, or jumps if the price settles to a new level. Also in this case, the amplitude and frequency of these large variations are not uniformly distributed in time. In order to better visualize these two observations, in Fig. 5.3, we reported a time-dependent empirical estimation of volatility, based on the estimation described in Refs. [34, 35].

---

[d]In this context, the aspect to be projected into the future is defined by the risk-factor variation.

The two effects outlined above are in contrast with the i.i.d. assumption underlying the historical method that seems to require some care for its application. For example, a typical solution to the heteroskedastic behavior of volatility is to appropriately reduce the sample length in order to take into consideration only realizations closer to the last available price. In this way, given that the volatility aggregates itself in clusters, one could expect that defining in a suitable way the length of the cluster, one could select only the last cluster of volatility. In this way, the scenario generation would be coherent with the last level of volatility that, because of clustering, should be a good approximation for the next future levels of volatility. On the other side, this sample reduction would imply a larger statistical error that should be appropriately managed. In general, sample length ranges between 250 and 500 daily observations.

Another solution to heteroskedastic volatility is a normalization procedure that drops volatility time dependency in the variation time series. Typically, one assumes that volatility *proportionally* affects the amplitude of the variations, so one can cancel the heteroskedastic effect simply dividing each variation for a suitable estimation of local volatility. We refer to *local* volatility in order to stress that our estimation has to refer to a local property of the risk-factor variations in order to obtain the volatility time dependency [36]. In formulas, assuming that for each time $\{t_1, \dots, t_{n-1}\}$, an estimation of the local volatility is provided $\{\sigma(t_1), \dots, \sigma(t_{n-1})\}$, one can adjust the risk-factor variations to the last available volatility level $\sigma(t_{n-1})$, considering the following renormalization:

$$\Delta S(\tilde{t}_j) \equiv \frac{\Delta S(t_j)}{\sigma(t_j)} \sigma(t_{n-1}) \quad \forall j = 1, \dots, n-1. \tag{6.44}$$

With this approach, one can transform the variations in a more homogeneous time series, as it can be observed comparing Fig. 6.5 with its normalized version.

As one could argue, the efficiency of renormalization strongly depends on the accuracy of the local volatility estimator. The main drawbacks of this method are as follows:

- Typically local volatility estimation is affected by a large statistical uncertainty as, in order to be *local*, it is based on a few observations in most cases.
- Local volatility algorithms are quite sophisticated and, in general, could slow down risk computations, when the number of risk factors is large.

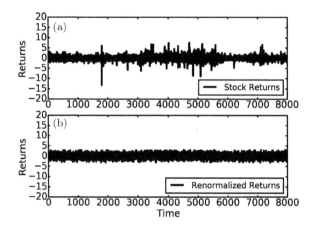

Fig. 6.5. We show (a) the historical series of IBM relative returns (already shown in Fig. 5.3) and (b) its renormalized version with the exclusion of outliers. It is evident that time series in (b) is much more homogeneous than (a).

- Renormalization is based on an ex-post adjustment of the dataset, so it could artificially misrepresent the information implied by the dataset itself. This can be an issue, especially from a regulatory point of view.

These considerations conclude our discussion about the historical approach of the estimation of future risk-factor distribution. By simple examples, we showed the strengths and the weaknesses of this method and how to deal with them. In the following sections, we will show alternative methods for the estimation of risk-factor distribution and, in the same way, we will discuss their features. In any case, we can observe that none of them represents a definitive solution of the problem, and some adjustment will be required in order to deal with their weaknesses.

6.6.2. *Parametric Method*

In the previous section, we discussed about the historical method for the estimation of future risk-factor distributions. The main idea of this method is not to make any choice on the theoretical distribution of risk factor and to consider the empirical distribution of risk-factor variations as a good proxy for the desired distribution.

On the contrary, the parametric method requires a hypothesis about the theoretical (parametrized) distribution of the risk factor variations and uses historical data to find a proper parametrization for it. This approach completely changes our perspective as the distribution and in particular tail behavior of the risk factors is now postulated from the very beginning. As a consequence, one can use information embedded in the central part of the distribution, typically with low uncertainty, to calibrate the parametrized distribution and exploit this result to obtain information about low probability events. In particular, if the chosen distribution is analytically known, percentile estimation is straightforward, and its uncertainty depends only on the accuracy of the parameter estimation. For example, if one assumes that the theoretical distribution of the risk-factor variations were well represented by a Gaussian, it is sufficient to use standard mean and variance estimators to deduce the distribution, and then obtain the percentile at any confidence level with the usual Gaussian analytical formula.

Obviously, some drawbacks have to be mentioned. First of all, the choice of the theoretical distribution is of fundamental importance for the final results of the model. From an empirical point of view, it often results that the Gaussian assumption has poor results in fitting typical risk factors (as equity prices variations). In fact, typical price variation distributions exhibit *fat-tails*, i.e. the probability to observe large fluctuations is typically larger than the one that can be estimated by a Gaussian distribution. In Fig. 6.6, we show the empirical PDF and CDF obtained by the time

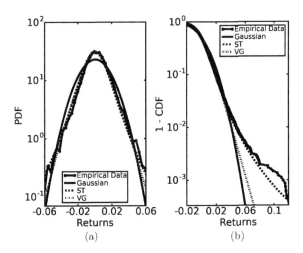

Fig. 6.6. We show (a) the empirical IBM relative returns PDF and the fit with Gaussian, Student's t and VG. In (b), we show the corresponding CDF results. It is evident that Gaussian fit has poor results with respect to the other two distributions. In this case, ST distribution is in better agreement with the empirical tail of the distribution (see snapshot (b)).

series of IBM relative price variations, considering more than 8000 daily observations. We show on a semi-log scale the fitting results considering different theoretical PDFs and CDFs. As stated above, the Gaussian PDF has poor fitting results especially in the tails, as magnified by the semi-log scale. This fact is not in contrast with the result of the central limit theorem because, as already outlined in Section 2.8, its effectiveness is assured only in the central part of the distribution. This empirical evidence is quite robust across different price variation distributions, and it poses serious problems from a risk management perspective. In fact, as already mentioned in Section 6.3, risk is essentially a matter of large and rare event estimation that relies on tail characterization. In this respect, the Gaussian approximation would lead to an underestimation of the probability of risky events, i.e. to the underestimation of the risk. From this perspective, we observe that the historical method, described in Section 6.6.1, is not affected by these kinds of problems as it naturally relies on the empirical distribution of the risk factors, so the tail behavior is coherently taken into account.

A natural solution to this important problem is to consider theoretical distributions that have a decay slower than Gaussian, i.e. $\exp(-x^2/2)$. In this respect, firstly it is worthwhile to clarify some aspect of the fat-tailed distributions. In general, by *fat-tailed* distribution, one refers to distributions that exhibit a power-law decay in the CDF, i.e $F(x) \sim |x|^{-\alpha}$ for $x \to \pm\infty$. In this book, we make use of slight abuse of terminology, and we will associate the fat-tail terminology to each distribution showing a decay slower than the Gaussian one, including exponential-like distributions, with PDF characterized by $\exp(-\alpha|x|)$ for $x \to \pm\infty$. In addition, we observe that this definition refers to very large (or very small) values of x. From a practical point of view, this requires a definition of a probability level above (below) which the fat-tailed behavior is evident as the x value is large (small) enough. As an example, we show in Fig. 6.7 a comparison between Gaussian and Student's t-distribution with the same standard deviation. By this example, it is evident that the fat-tailed behavior is present only for extremely low (high) probability level (less than 2.5%). In particular, it is not true that a power-law decaying distribution always implies a higher probability for large fluctuations of the risk factors, but it depends if the considered probability level is large (or small) enough. This aspect could lead to counterintuitive behaviors in practical situations, where, in extreme cases, a Gaussian distribution would imply a larger VaR measure than a power-law distribution. For example, in Fig. 6.7, it is evident

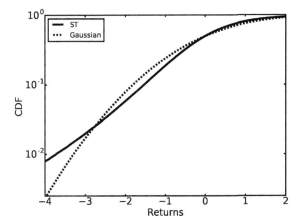

Fig. 6.7. We show the CDF for a ST with 4 degrees of freedom and a Gaussian distribution with the same variance. The fat-tails behavior of the ST distribution is evident for a probability less than 2.5% where the straight line is over the dotted line. For larger confidence level, the Gaussian distribution could imply a larger risk estimation.

that at confidence level 10%, a Gaussian distribution implies an loss larger than an ST distribution. Anyway, in most of the cases for risk estimation the following relation holds:

$$\text{Risk}_{\text{power-law}} > \text{Risk}_{\text{exponential-like}} > \text{Risk}_{\text{Gaussian}}. \qquad (6.45)$$

After this discussion on the main aspects of the fat-tailed distributions, we can come back to our original problem on the choice of the theoretical distribution for risk-factor variations. As the Gaussian distribution shows poor fitting results, one typically relies on exponential-like or power-law distribution families. The tail behavior of the distribution is the main driver in the risk estimation, so the choice between these two families is a crucial aspect of the parametric method. Unfortunately, from an empirical point of view, it is very difficult to discern between these two families for the two following reasons:

- In order to analyze a power-law behavior, qualitatively one needs to observe variations (in tails) of, at least, two orders of magnitude over the two axes, namely the risk-factor variations and the probability axes. This requires quite a large amount of data that could span quite a large region of extreme events. On the contrary, if the region spanned by the tail of the distribution is not large enough, it is difficult to discern between the two distributions.

- In order to avoid mixing different volatility clusters, one typically focuses on the observations in the dataset closer to the reference date, reducing the length of the sample (see also discussion on heteroskedastic volatility in Section 6.6.1).

As the two previous points are in contrast, the optimal choice is still debated.

In general, in order to gain some insight on this issue, we considered the following experiment:

- We considered a large sample of different equity price time series, each composed by more than 5000 observations.
- We assumed that the relative variations of the prices can be a suitable example of risk-factor variations.
- We considered the Student's t-distribution (ST) and Variance-Gamma (VG) distributions as representative of the power-law and exponential families, respectively.
- We fitted each empirical price variation distribution with the considered theoretical distributions, and we compared the distributions of the fitting errors in the two cases.

In Fig. 6.8, we report the fitting errors distribution of the two considered distributions. From the figure, it is evident that there is not a clear winner between the two distributions as fitting performance depends on the considered time series. For example, in the IBM case, the ST distribution has much better fitting performances than VG as it can be observed in Fig. 6.6, but this behavior cannot be generalized to all the considered historical series.

To conclude our discussion about the drawbacks of the parametric method, we stress that when risk related to a financial instrument depends on several risk factors, the correlation structure should be estimated and taken into consideration. In this case, a multi-variate representation is required with additional complications for the covariance estimation.

From our discussion, it is evident that the parametric method embeds both positive and negative features. Typically, in order to develop and actually use a model for risk estimation, careful and detailed analysis is performed in order to leverage on the strengths of the model and address all the issues.

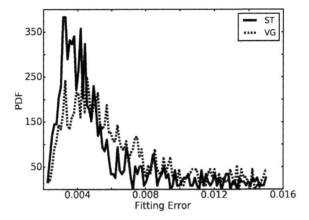

Fig. 6.8. We show the fitting errors obtained repeating the fit of the empirical distribution of about 2000 time series of relative returns, obtained considering main S&P500 index components. In particular, we considered VG and ST distributions. Even if the fitting errors of the Student's *t*-distribution show a smaller variance, the fitting performances of the two distributions are overall comparable and a clear winner cannot be identified. In our experiment, we also tested the fitting performance of the Gaussian distribution that was one order of magnitude worse and we did not report this in the figure.

6.6.3. *Monte Carlo Method*

The last method we are going to discuss for risk estimation purposes is the so-called Monte Carlo (MC) method. This method is based on the Monte Carlo simulation technique that will not be discussed in this book. The interested reader can find details in Refs. [37, 38]. In a nutshell, the MC technique in QF is a numerical method that generates random numbers (sampling from a given theoretical distribution) and uses them to simulate stochastic paths of a given Stochastic Differential Equation (SDE). So, with respect to risk estimation, the approach is very straightforward as one can generate risk factor scenarios and estimate what the portfolio variation in each scenario is. This approach is interesting as one can work directly on the *dynamics* of the risk factor, possibly comparing it with the empirical one. In principle, one is free to specify the time dependency and the stochastic nature of the drift and the diffusion part, adding also spikes and jumping behavior and simulating numerically a very complex SDE. From this perspective, one is in the position to take into account effects of market movements like heteroskedastic volatility that are hard to be modeled with other approaches.

Also in this case, some drawback is implicit in this methodology and should be coherently managed by risk Quants. As for the parametric method, when different risk factors are contextually taken into consideration, a covariance matrix should be estimated and stored in the computer memory introducing potential issues on the performance. On the same lines, typically, the more complex the stochastic model, the higher the computational effort required to generate scenarios for the risk computation. As a consequence, in most of the cases, the modeling freedom is reduced because of technological constraints. Finally, one should take into consideration that typically a complex SDE requires a non-trivial calibration procedure. In the worst cases, the statistical uncertainty of model parameters could generate results less robust and with performances poorer than the ones obtained by considering simple approaches.

6.7.   Does it Really Work? The Backtesting Approach

As already discussed in this book, modeling activity requires, by definition, the description of a complex system by simplifying hypotheses that represent the most important features of the system itself, avoiding the inclusion of negligible details. As a consequence, the resulting model is subjected to approximations that we deem do not affect the overall reliability of the results. On the other hand, during our discussion about risk modeling (and pricing model as well), we introduced many hypotheses that in the end allowed us to obtain quantitative estimations of the risk.

At this point, a very natural question is: *do these models really work? Are their risk measures in line with the actual risk we are going to face in the future?* In our opinion, the answer to this question is a fundamental part of the modeling activity, and it cannot be neglected. After all, quants are paid to tell management how risky an activity is and not to discover new formulas or new mathematical framework!

From a more theoretical point of view, model verification is, in our opinion, related to the so-called *scientific method* that requires each model results to be verified by consistent empirical experiment. Unfortunately, this framework hardly applies to QF world because typically only one time series can be used and, as a consequence, only one observation is available for each fixed time. In any case, we could at least verify if, applying our model as of a past date, our forecasts are in good agreement with the realized history. This is exactly the idea of the *backtesting* approach.

In order to clarify the idea, we assume that a 10-year historical series of an equity closing prices[e] $\{S(t_1), \ldots, S(t_m)\}$ is available (where m is the total number of available prices), and we have already developed a VaR historical risk model based on $n \ll m$ observations that can make forecasts about riskiness. In particular, we assume that our portfolio is composed by a given amount $A(t)$ of the equity, so that the only relevant risk factor is given by the equity price itself, and the pricing function to map risk factor variations into P&L is simply a linear function that multiplies the variations by the amount of equity. Despite our simplifying hypotheses, this setup can be easily extended to more general situations for VaR backtesting.

In order to backtest our model, we consider MR estimations as of past dates, and we compare them with the realized P&L variations. For example, assuming that our VaR time horizon Δt is one day, i.e. we are making forecasts about potential losses that could happen in one day, we use the subset $\{S(t_1), \ldots, S(t_n)\}$ to historically estimate the VaR (at a given percentile level α) that refers to time $t_{n+\Delta t} = t_{n+1}$, i.e. we are forecasting the P&L distribution at time t_{n+1} given the previous n observations. As we are backtesting as of past date, the price $S(t_{n+1})$ is already available in our historical series, so we can compare our risk estimation with the P&L that *would* be realized between t_n and t_{n+1} *if we had kept constant our position over the chosen time horizon*, i.e. assuming that $A(t)$ is constant over the time horizon Δt. In other words, we are assuming that variations in the position on a time scale smaller that the VaR time horizon can be neglected. In practice, because of this assumption, the P&L used for backtesting purposes is typically different from the realized P&L, as it neglects all intraday operations; in the following, we will refer to it as *backtesting P&L*.

To summarize, in this framework, we have a set of VaR estimations $\mathrm{VaR}(t_{n+1}), \ldots, \mathrm{VaR}(t_m)$ and the respective backtesting P&Ls $\mathrm{P\&L}(t_{n+1}), \ldots, \mathrm{P\&L}(t_m)$ that should be compared in order to assess the robustness and the reliability of our risk model. Firstly, we observe that the equity-amount function $A(t)$ of our example is piecewise constant over each Δt, and it is deterministically given at the estimation time. For this reason, it simply acts as a scaling factor, and one can focus on the comparison

---

[e]The closing price is the final price at which a security is traded on a given trading day.

between the forecasted distribution of the price variations with the realized price variation. In addition, we observe that as VaR represents a forecast of future events, none of the future information used for backtesting P&L calculation must be included in VaR modeling.

In order to compare VaR and backtesting P&L, one can rely on the definition of VaR as the maximum loss within the given confidence level α, and interpret the backtesting as a "hit or miss" problem where, if our model is coherent, we have a probability α of experiencing a loss larger than our VaR estimation. As a consequence, one can generate a zero–one sequence of random variables, evaluating if we experienced a breach with respect to our VaR estimation, i.e.

$$Br \equiv \left\{ \mathbf{I}_{\text{VaR}(t_{n+1}) < \text{P\&L}(t_{n+1})}, \ldots, \mathbf{I}_{\text{VaR}(t_m) < \text{P\&L}(t_m)} \right\}, \tag{6.46}$$

and we can analyze the Br statistics. In particular, *assuming that each VaR and P&L realization is independent*, one can expect that the Br random variables are distributed as a binomial distribution $\text{Bin}(\omega, m - n, \alpha)$ (Section 2.7.2). In particular, one could expect that for large value of $m - n$, the number of breaches should converge to the expected value of the binomial distribution, i.e. $(m - n)\alpha$. In general, using this approach, one can compare that statistics of the realized breaches on a given period $m - n$ with the binomial distribution $\text{Bin}(\omega, m - n, \alpha)$, and test if the number of breaches is in line with expectations within a given confidence level ω.

As an example of application, we considered the Intel Corporation historical series of prices from 15/11/1982 to 12/09/2012, and we derived the VaR historical estimation, considering the relative variations of prices. In particular, we considered a 5% confidence interval, a VaR time horizon of 1 day, and time window of 500 days for VaR estimation. In this way, starting from a time series of 7525 price variations, we obtained $k = 7525 - 500 - 1 = 7024$ VaR estimations[f] and the corresponding price variations to be used to identify the VaR breaches, Br. Assuming that each VaR estimation is independent, one should expect about $7024 \times 0.05 = 351.2$ breaches in our sample. In Fig. 6.9, we show the result of our experiment where we obtained 331 breaches. Considering that the standard deviation of the Binomial distribution is given by $\sqrt{k\alpha(1 - \alpha)}$,

---

[f]The factor -1 is due to the fact that for the most recent day, we can obtain a VaR estimation, but it cannot be compared with the next day variation. For this reason, the last VaR estimation cannot be used for backtesting purposes.

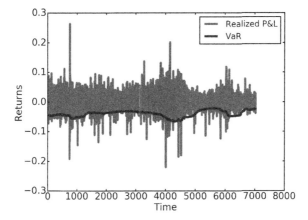

Fig. 6.9. We show the realized P&L variations and historical VaR estimation at 95% confidence level. In the considered case, we observed 331 VaR breaches over 7024 observations, which is slightly less than expected value, i.e. 351.

in our case, we have that the realized number of breaches lies about one standard deviation from the expected theoretical value, so the model performance can be considered statistically in line with expectations. On the other side, it can be observed that the breaches are not uniformly distributed with respect to time and on the contrary they seem to be organized in clusters. The reason for this behavior seems to be related to the "reactivity" of the model to a market change. In fact, as we are considering 500 observations for the VaR estimation, moving from one day to another one, we are changing only two observations (the oldest observation is removed, and it is substituted by the most recent) over 500 of our sample. As a consequence, when a new volatility cluster is present in our historical series, there is a transient time that is required by our risk model to adapt itself to the new regime. During that time, if the new volatility cluster has a higher variance, it is likely to observe recurrent breaches; this fact justifies their non-uniform distribution. From a statistical perspective, this behavior is a consequence of the overlapping nature of the time window considered for VaR estimation that would imply that our VaR estimations are not independent.

Repeating this backtesting approach over a large number of historical series, it can be observed that the number of realized breaches is quite different from the expected one, showing that some adjustments in our risk forecasting approach should be considered in order to improve performances. As an example, in Fig. 6.10, we show the distribution of the

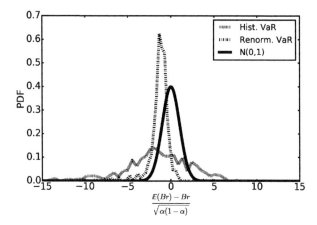

Fig. 6.10.   We show the results of our backtesting experiment considering the distribution of performance parameter of Eq. (6.47) for standard historical VaR estimation, the renormalized VaR and the theoretical distribution of the performance parameter (approximated here by a Gaussian of zero mean and variance equal to one).

coefficient of variation CoV of the breaches,

$$\text{CoV} = \frac{Br - \mathbb{E}(Br)}{\sqrt{\alpha(1 - \alpha)}} \tag{6.47}$$

for a large number of time series, considering the following VaR models:

- Historical VaR model — It is the model already discussed above and in Section 6.6.1.
- Historical Locally Renormalized VaR model — Historical VaR model where the risk-factor variations are renormalized, considering a local volatility estimator [34].

From the figure, it seems that the renormalization technique improves the backtesting results, reducing the breach fluctuations and centering the distribution around the expected value. In any case, it can be observed that both the models show on average a larger number of breaches than expected. This undesirable effect is a consequence of the hypothesis of independent returns that is typically not satisfied in equity market. On the contrary, it can be observed by historical series analysis that abnormal jumps are typically correlated in time and this fact slightly increases the number of breaches.

    As already stated above, this backtesting methodology can easily be extended to more complex model setups, and it represents quite a standard market practice. Recently, new backtesting methodologies were proposed [39–41] in order to obtain a better indicator of the model performance, mainly focused on the entire forecasted distribution of the P&L instead of concentrating only on a percentile level.

In addition, we observe that the discussed backtesting methodology is well suited when VaR time horizon is small enough (i.e. 1 day) to allow the estimation of statistically meaningful tests. On the contrary, when time horizon increases, the required time series length to obtain statistically meaningful results also increases. For example, in order to test a 10-days time horizon VaR with 250 independent observations, one needs a time series of 10 years; while for 1-day time horizon VaR, a time series of 1 year can be used. For this reason, in order to test long time horizons, one typically considers tests on time scaling relations between $1-$ VaR and longer time horizons VaR.

6.8. From Theory to Practice

In the preceding sections, we discussed a theoretical framework for risk estimation and we proposed practical examples of application. In addition, we discussed how to test our models using empirical evidences. Even if this framework is, in our opinion, a very good starting point to understand practical QF problems and how mathematics can be efficiently exploited to solve them, it is a rather simplified version of how risk models are managed in large financial institutions.

For the sake of completeness, in this section, we describe how risk models are typically managed in a bank in order to give a general intuition of the complexities that must be tackled in practical situations. In order to reduce the scope of our discussion, we will focus on the management of the trading booking MR.

At the time of writing, regulatory requirements (Basel III [42], Capital Requirements Regulation (CRR) [43] and Directive (CRD IV) [44]) for the estimation of the MR are quite wide and are based on the estimation on the following components of the MR:

- **Value-at-Risk (VaR)**: As described above in detail, it estimates, at a given confidence level (typically 99%), the potential decline in the value of a position or a portfolio under normal market conditions using methodologies like the ones described in Section 6.6.

- **Stressed Value-at-Risk (SVaR)**: It is the same risk measure mentioned above, but estimated under extreme stressed scenarios that typically consider systemic and business-specific stresses. Systemic stresses are usually designed to quantify the potential impact of extreme market movements, while business-specific stresses are introduced to probe the risks of particular portfolios and market segments, especially those risks that are not fully captured in standard VaR measure.
- **Incremental Risk Charge (IRC)**: It represents a charge to cover the default and credit migration risks of non-securitized credit products. In other words, it measures the loss that a bank can undergo because of the decline of the credit worthiness of bond (or other financial instruments like Credit Default Swaps (CDS)) issuers. For example, if a bond issuer migrates from a high credit rating to a lower one, the probability of receiving back the invested money at bond maturity decreases. As a consequence, the value of the bond issued will decrease.[g] Incremental Risk Charge measures this kind of risk. Typically, it is measured over 1-year time horizon at a 99.9% confidence level under the assumption of constant positions (i.e. the positions are assumed to be constant over 1 year).
- **Comprehensive Risk Measure (CRM)**: A formal definition can be *an estimate of risk in the correlation trading portfolio, taking into account credit spread, correlation, basis, recovery and default risks.* In other words, CRM can be understood as an IRC-type charge where additional risk factors must be included (e.g. the volatility of the correlation between financial instruments, multiple defaults correlation, etc.). We observe that this kind of risk is subjected to a floor imposed by regulations.
- **Other Charges (OC)**: Securitization Charges, Standard Specific Risk Charges, Capital Charges are not included in the VaR model.

In particular, regulations require that all the components mentioned above must be taken into consideration for risk estimation and they must be summed up considering suitable multipliers for each component that depends on the model used and, for what concerns VaR and SVaR, on the number of breaches in VaR backtesting. This sum represents the capital charges required for trading books.

---

[g]We observe that this effect is not reflected in Eq. (3.20), where we neglected default probability.

To conclude this section, it must be mentioned that the requirements for capital are going to change in the future as the result of the so-called *Fundamental Review of the Trading Book* (FRTB) [45]. It represents a set of proposals by Basel Committee on Banking Supervision (BCBS) for the future framework of MR regulatory capital rules for banks. The scope of new proposals is quite wide and it is out of the scope of this book. We only mention that two key points are related to the introduction of the *Expected Shortfall* (Section 6.4) as risk measure and the estimation of the *Default Risk Charge*, instead of IRC, in order to consider the impact of the default on the trading books.

Chapter 7

The New Post-Crisis Paradigms

7.1. The Financial World After Financial Crisis

In the recent years, financial markets were affected by a global financial crisis that significantly influenced the fundamental paradigms of options pricing and risk estimation. In this respect, an interesting description of what happened during recent years can be obtained considering the evolution in time of the difference (spread) of two main interest rates: Euribor and Eonia rates. Following the detailed analysis in Refs. [46, 47], it is possible to define as follows:

- Euribor: *the reference rate for over-the-counter (OTC) transactions in the Euro area. It is defined as the rate at which Euro interbank deposits are being offered within the EMU zone by one prime bank to another at 11:00 a.m. Brussels time. The rate fixings for a strip of 15 maturities, ranging from one day to one year, are constructed as the trimmed average of the individual fixings (excluding the highest and lowest 15% tails) submitted by a panel of banks. The Contribution Panel is composed, as of September 2010, by 42 banks, selected among the EU banks with the highest business volume and credit standing in the Euro zone money markets, plus some large international banks from non-EU countries with important euro zone operations. Thus, Euribor rates reflect the average cost of funding of EU banks in the EUR interbank market at each given maturity.*
- Eonia: *the reference rate for overnight OTC transactions in the Euro area. It is constructed as the average rate of the overnight transactions (one day maturity deposits) executed during a given business day by a panel of banks on the interbank money market, weighted by the corresponding transaction volumes. The Eonia Contribution Panel*

coincides with the Euribor Contribution Panel. Thus, Eonia rate includes information on the short-term (overnight) liquidity expectations of banks in the Euro money market. It is also used by the European Central Bank (ECB) as a method of effecting and observing the transmission of its monetary policy actions. Furthermore, the daily tenor of the Eonia rate makes negligible the credit and liquidity risks reflected on it: for this reason, the OIS rates are considered as the best proxies available in the market for the risk-free rate.

In other words, the difference between Euribor and Eonia rates can be considered as a good proxy about how credit and liquidity risk is perceived by market participants. In particular, it is possible to split the history of the financial crisis in four different periods (see Fig. 7.1):

- Before August 2007: The difference of the two rates is negligible, so the credit and liquidity risk premia are not to be taken into account for modeling purposes.
- Between August 2007 and March 2009: The spread increases remarkably, and it hits its peak in October 2008 after Lehman Brothers default.

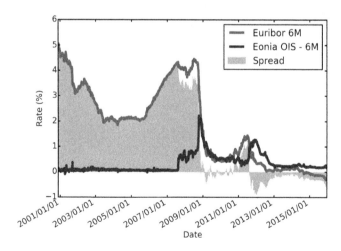

Fig. 7.1. We show the time series of the Euribor and Eonia rates for a fixed tenor (6 months) and the difference between the two rates (spread). It can be observed that before 2007, the spread between the two rates was negligible, while after August 2007, it became positive. In addition, it can be observed that after the end of 2014, both the two rates became negative. This fact had a relevant impact on the modeling assumptions of the interest-rate dynamics.

- Between March 2009 and mid-2010: The spread decreases and stabilizes at the non-negligible level of about 40 bps. This effect was probably due to the Central Bank policy that decided to provide banks a large amount of liquidity in order to control the effects of the crisis and restore the credit confidence within interbank market.
- Between mid-2010 and end-2011: The market interest-rate dynamics was mainly influenced by the possible consequences related to possible failure of some European states in the Euro financial system. In this context, the ECB injected liquidity in the market allowing banks to fund themselves on very low interest rates. As a consequence, Eonia decreased remarkably implying a reduction of the Euribor rate.
- After end-2014, both the two rates became negative. This fact had a relevant impact on the modeling assumptions of the interest-rate dynamics, in particular, models based on lognormal distribution, implying a positive interest rate required to be corrected in order to take this feature into account.

Given this financial context, many old paradigms on which we based our theoretical framework appear not to hold any more, and different *adjustments* were required in order to obtain a reliable description of the new financial system. Unfortunately, some of these adjustments are quite technical and they would require a practitioner's knowledge of financial contracts that is out of the scope of this book. For this reason, in the following sections, we provide a very simple description of these new technical aspects, giving a glimpse of the mathematical approach that should be followed, in order to keep these new features into consideration. We will refer to more specific literature for more detailed descriptions.

7.2. Multi-Curve Framework

In this section, we discuss how the theoretical pricing framework, described in Chapter 4, has been changed as a consequence of the financial crisis. In particular, in Section 7.1, we remarked that after 2007, new sources of risk (liquidity and credit risk) were included in the interest rates traded in the market, that required, as a consequence, different risk premia. For this reason, differences in interest rates, which before August 2007 were considered negligible, became substantial and required the introduction of an *ad hoc* model. In particular, nowadays it is not possible anymore to mix different interest rates referring to different contract maturities in order to define a unique reference curve as a numeraire of our pricing routines.

In order to better explain the situation, we reconsider as an example the replica approach of Section 3.3.4.4 for the pricing at time t of a forward rate agreement $(\mathrm{FRA}(t, S, T))$ that fixes at time S and with maturity T. In this case, we considered a replica strategy based on going long and short on different bonds with maturities S and T with a notional chosen in order to replicate the final FRA payoff. In Section 3.3.4.4, we claimed that as this strategy provides the same payoff of the FRA, its price must be equivalent to FRA price. We implicitly assumed the following:

- The issuer we are going to lend money (bond) cannot default.
- There are no liquidity issues at time S that could prevent us to swap from the bond with maturity S to the one that starts at S and has maturity equal to T.

Obviously, if these two hypotheses were not satisfied, we could not claim that the FRA price is equivalent to the strategy, as the latter would imply a higher risk than an FRA purchase. As a consequence, the FRA price should be higher than the replica, as it actually avoids liquidity and default issues. This is exactly the effect we observed on the market after the financial crisis, when market participants started to perceive liquidity and credit risk also for this kind of products, and started to require a risk premium. As a consequence, in order to correctly reproduce market prices, one should extend the theoretical pricing framework we discussed in Chapter 4, adding a suitable and comprehensive model for liquidity and default risk of all the contracts available in the market.

Unfortunately, such a general framework is still not available in a truly comprehensive formulation, and market practice seems to follow a different approach that is known as *multi-curve framework*. The multi-curve approach is based on the idea of considering the rate length (tenor) as the main driver to describe the above-mentioned rate discrepancies. In particular, interest rates that belong to the same *tenor class* require a similar risk premium for credit and liquidity issues. As a consequence, in order to take into consideration post-crisis effects on interest rate pricing models, one needs to treat each tenor class as instruments affected by the same level of riskiness; the different risk classes cannot be mixed together in a unified framework. As a consequence, consistency relations, similar to the one mentioned above between FRA and bonds instruments, can be found only inside the same tenor class, where the risk premium related to credit and liquidity issues can be considered comparable.

A peculiar aspect of the multi-curve framework is that the modeling assumptions made to describe the discount factor dynamics, or in general, the tradable asset we take as numeraire, have to take into consideration the tenor class we consider to model the risk-free rates and their relation with the payoff of the instrument we want to price. As an example, we consider again the problem of the estimation of the expectation of the forward rate $F(t, S, T)$ in the measure obtained taking the bond $P(t, T)$ as numeraire. In Section 5.6.1, we obtained that *as the bond we took as numeraire has the same tenor of the one in the forward definition*, Eq. (5.48), the forward rate is a martingale with respect to the $P(t, T)$-bond measure (Eq. (5.52)). This result is based on the fact that, if the multi-curve framework is not considered, the forward rate multiplied by the bond $P(t, T)$ is a tradable asset as it represents the difference of two bonds. On the contrary, if the multi-curve framework is taken into consideration, this relation is not assured to hold anymore, as the segmentation of the interest rates by tenor class cannot be neglected.

In order to understand this fact, we assume that for each maturity T of interest, a fictitious bond $P_D(t, T)$ that represents the expected value of the discount factor is defined, as an extension of Eq. (3.20) for stochastic risk-free rates,

$$P_D(t, T) \equiv \mathbb{E}^{\mathbb{Q}} \left(e^{-\int_t^T r(u) \mathrm{d}u} \middle| \mathcal{F}_t \right). \tag{7.1}$$

The subscript D stands for discount factor, and it reminds us that, in order to define this bond, we referred to the risk-free rates $r(t)$ which belong to a specific class where we think that all the risks are negligible. In practical situations, one has to discount payoffs at different maturities, so Eq. (7.1) should be regarded as a *curve* definition (as a function of T) more than a single bond definition. On the same lines, we define the forward rate, making explicit the tenor class the forward belongs to

$$F_x(t, S, S + x) \equiv \frac{1}{x} \left(\frac{P_x(t, S + x)}{P_x(t, S)} - 1 \right), \tag{7.2}$$

where x labels the tenor class we are considering. Given these new definitions for the discounting bond curve and the forward rate, one could try to apply the same reasoning used in Section 5.6.1 to estimate the expected value of the forward rate. Unfortunately, in this case, notation helps us to remember that in general, the bonds implied in the forward definition P_x

and the bond of the discounting curve P_D can be different in principle. As a consequence, it is not assured that the product $F_x(t, S, S+x)P_D(t, S+x)$ is a tradable asset and, lastly, that the forward rate is not a martingale in the P_D measure

$$\mathbb{E}^{\mathbb{Q}_{P_D(t,S+x)}}\left(F_x(S, S, S+x)|\mathcal{F}_t\right)F_x(t, S, S+x). \tag{7.3}$$

As a consequence, one needs to introduce a new curve (a function of S) to describe the expected value of the forward rate for *each* tenor class x under the measure $\mathbb{Q}_{P_D(t,S+x)}$; this justifies the term *multi-curve* to refer to this framework. In particular, one usually refers to the discounting curve $P_D(t, T)$ as the curve used to discount all the cash flows related to the contract and the forwarding curve that describes the expected value of the forward rate $F_x(t, S, S+x)$. In this respect, it is possible to prove that the expected value of the forward rate is equal to FRA rate that we defined in Section 3.3.4.4 as the rate that makes the price of the FRA equal to zero. As a consequence, a typical approach is to use the FRA rates to build the forwarding curve, exploiting the relation,

$$\text{FRA}_x(t, S, S+x) = \mathbb{E}^{\mathbb{Q}_{P_D(t,S+x)}}\left(F_x(S, S, S+x)|\mathcal{F}_t\right) \tag{7.4}$$

that allows the generation of the forwarding curves for all the different tenors x available in the FRA market.[a] In general, we can summarize the new pricing framework slightly modifying the risk-neutral pricing formula, making explicit the multi-curve setup as

$$V(t) = P_D(t, T)\mathbb{E}^{\mathbb{Q}_{P_D(t,T)}}\left(V(F_x(T))|\mathcal{F}_t\right), \tag{7.5}$$

where $V(t)$ is the value of a contingent claim at time t and $F_x(T)$ represents a generic underlying at time T (or a vector of underlyings) that insists on the tenor x. In order to correctly apply this generalization of the risk-neutral formula, one needs to estimate the forwarding curve $\mathbb{E}^{\mathbb{Q}_{P_D(t,T)}}\left(F_x(T))|\mathcal{F}_t\right)$ as a function of T, typically relying on calibration procedures and market data.

To conclude this section, we remark that as in the case of the forwarding curve, different choices are also available in the case of the discounting curve, for example as a function of the collateral contractual agreement underlying an instrument. In fact, it is a market practice to discount the cash flows generated by an instrument with different discounting curves, taking into consideration the amount and the quality of the collateral

---

[a]For large maturities, special instruments, the *swaps*, are used instead of the FRAs.

(if any) exchanged by the counterparties to hedge the counterparty credit risk. Further details on this interesting topic can be found in Refs. [11, 12, 48].

7.3. Fair Value Adjustments

As a consequence of recent financial crisis, the general theoretical framework about rational pricing resulted in being weakened by commonly accepted hypotheses that appeared to be wrong. In particular, the credit and liquidity risks, which before August 2007 were not taken into consideration for main interest rate contracts, cannot be neglected anymore (see discussion in Section 7.1) and needed to be included in the theory of option pricing. In other words, after financial crisis, market participants started to *price* new sources of risk requiring suitable adjustments to the price. If at the very beginning of this new revolution, the main risk to be priced was related to the risk of default [generating the so-called Credit Value Adjustment (CVA)] many other sources of uncertainty were required to be priced: Debt Value Adjustment (DVA), Funding Value Adjustment (FVA) and so on.

For this reason, nowadays practitioners generically refer to xVA, where x represents a generic unknown source of valuation adjustment.

In addition, we observe that these adjustments are also linked to a set of accounting principles [49] that explicitly mention them for the fair value definition. This aspect increases somehow the relevance of this adjustment also for accounting purposes and justifies the big number of papers recently published on this topic. In the following sections, we will focus on CVA, and we will introduce the simple mathematical aspects in order to deal with this kind of adjustment. The following discussion should be regarded as a simple introduction to the topic and it is by no means exhaustive.

7.3.1. *Credit Value Adjustment*

CVA represents a modification to the fair value due to the required premium because of default risk of the counterpart. This premium can be estimated taking into consideration the main risk factors that would drive the amount of loss in the case of the counterpart default. As a consequence, the price of financial instruments must be modified in order to take into consideration default risk, as given by the formula,

$$V(t) = \mathbb{E}^{\mathbb{Q}}(D(t,T)V(T)|\mathcal{F}(t)) - \text{CVA}(t), \qquad (7.6)$$

where $\text{CVA}(t)$ represents the CVA at time t and the expectation on the right-hand side of the equation is the usual risk-neutral formula, where the CVA is neglected. As expected, the value of a financial instrument subject to counterparty risk is lower than the equivalent *default-free* instruments.

In order to estimate CVA, the first observation we should keep in mind is that the amount of our loss will depend on our exposure towards the counterpart at the default time. On the contrary, if at default time, we had a liability towards the counterpart, the loss is zero, as the counterpart default would not imply any loss from our perspective. To summarize these two cases, we can express our exposure at default (EAD) as the positive part of the instrument value at default time τ, $\max(V(\tau), 0)$. The second ingredient we need for obtaining CVA is the already mentioned default time τ. In particular, one is interested if the default time happens before or after the maturity of the contract. Because of this reason, one considers the indicator function $\mathbb{I}_{\tau < T}$ where T is the maturity of the contract.

When a default happens, creditors have typically the opportunity to recover only part of their credit that usually represents a percentage of the EAD. This percentage is usually called *recovery rate*, R. On the contrary, the lost percentage is called *Loss-Given-Default*, L_{GD}. The two parameters are obviously linked by the relation $L_{GD} = 1 - R$.

Given these main drivers, we can interpret the problem of the CVA estimation as the problem of pricing a (missing) cash flow at the (random) default time τ. As a consequence, we can exploit the standard risk-neutral pricing formula in order to estimate CVA,

$$\text{CVA}(t = 0) = \text{CVA}_0 = L_{GD}\mathbb{E}^{\mathbb{Q}}[D(0,\tau)\max(V(\tau),0)\mathbb{I}_{\tau < T}]$$

$$= (1 - R)\mathbb{E}^{\mathbb{Q}}[D(0,\tau)\max(V(\tau),0)\mathbb{I}_{\tau < T}], \qquad (7.7)$$

where $D(0,\tau)$ is the discount factor between $t = 0$ and the default time, $\mathbb{Q}$ is the usual risk-neutral measure, and we assumed that the Loss-Given-Default or, equivalently, the recovery rate is known and constant. In addition , for the sake of simplicity, we considered the CVA estimation at time $t = 0$. If we analyze Eq. (7.7) in detail, we can observe that the CVA pricing can be much more complex than the usual pricing problems because of the presence of the nonlinear max operator. This would increase the complexity of the pricing problem also for standard linear products for which analytic

formulas can easily be derived. In addition, one has to take into consideration also the term $\mathbb{I}_{\tau < T}$ that is typically modeled as an additional random
variable with a distribution that needs to be calibrated.

Alternatively, using Dirac's delta function property,

$$f(x)\mathbb{I}_{x<y} = \int_0^y f(u)\delta(u-x)du, \tag{7.8}$$

where $f(x)$ is a generic function and $\delta(x)$ is the Dirac's delta function, one
can express Eq. (7.9) as

$$\text{CVA}_0 = (1-R)\mathbb{E}^{\mathbb{Q}}\left[\int_0^T D(0,t)\max(V(t),0)\delta(\tau-t)dt\right]. \tag{7.9}$$

In order to simplify the problem and to obtain a reliable CVA estimation,
we observe that Eq. (7.7) can be written as

$$\text{CVA}_0 = (1-R)\int_0^T \mathbb{E}^{\mathbb{Q}}[D(0,t)\max(V(t),0)|t=\tau]p(t)dt, \tag{7.10}$$

where on the right-hand side, we conditionar the expectation of the EAD to
the information available at default time, and we integrated over the PDF
of the default time $p(t=\tau)$.

At this point, a typical choice is to assume that the default time is independent of the discounted EAD part, so that we do not need to condition
$t = \tau$ in the first expectation[b] on the right-hand side of Eq. (7.10). Denoting
the default probability with

$$\mathbb{D}(\tau < T) \equiv \mathbb{E}^{\mathbb{Q}}(\mathbb{I}_{\tau<T}), \tag{7.11}$$

or in differential terms,

$$dD(\tau < u) \equiv \mathbb{E}^{\mathbb{Q}}(\mathbb{I}_{\tau\in[u,u+du]})$$
$$= \mathbb{E}^{\mathbb{Q}}(\delta(\tau-u)du), \tag{7.12}$$

where we exploited Dirac's delta representation of Eq. (7.8), we can write
Eq. (7.10) by integral notation,

$$\text{CVA}(t=0) = (1-R)\int_0^T \mathbb{E}^{\mathbb{Q}}[D(0,u)\max(V(u),0)]dD(\tau < u). \tag{7.13}$$

Equation (7.13) is the most used expression to estimate CVA.

---

[b]Remember that two variables are independent when their expected value is not affected
by the information generated by the other variable.

This equation has a central role in CVA estimation as it allows one to distinguish between a pricing problem of the product $V(t)$ affected by optionality because of the presence of the *max* operator (first expectation in Eq. (7.13)) and the default probability part related to the second expectation. So, if the independence assumption holds, we can conclude that the CVA is actually a call option on the contract we want to price with a maturity equal to the default time and weighted by the default probability. We observe that in this framework, we only take into consideration the counterpart default probability, neglecting our probability of default. In this case, CVA is usually denoted as *unilateral* CVA.

As was already mentioned, the first part of the formula refers to a sort of optionality (embedded in the contract) due to the credit risk and it can be estimated by usual risk-neutral pricing approaches. In addition, we need to estimate the default probability of the issuer of the contract *under the risk-neutral measure*. Generally, this is not a simple task and it requires some care. Usually, it is a market practice to exploit the information on default embedded in particular contract called Credit Default Swap (CDS) from which it is possible to deduce, under suitable hypothesis, the risk-neutral probability function.

From a computational point of view, the most demanding part of CVA is related to the option estimation, essentially for two reasons:

- The payoff $\max(V(t), 0)$ can become really complex also for simple linear instruments, and an analytic formula could not be made available for most of the instruments in the portfolio. As a consequence, intensive Monte Carlo (MC) simulations are usually required to obtain a suitable estimation of the optionality part of the CVA.
- CVA is referred at counterpart level that typically requires the CVA estimation of *many* financial instruments aggregated in a proper and contractual way. As a consequence, when a new contract with a given counterpart has to be priced, the whole portfolio referring to that counterpart must be priced in order to understand the effects at CVA level.

For these reasons, in CVA estimation, the technological aspects have a relevant role and in most of the cases, smart solutions must be found in order to speed up the calculations.

In this respect, an interesting approach can be found in Ref. [50] where an approximated CVA expression is exploited in order to obtain a computationally efficient estimation.

The idea is based on the fact that the option part of the CVA can be decomposed into the product of the value of the underlying instrument $V(t)$ and the indicator function $\mathbb{I}_{V(t)>0}$, i.e.

$$\max(V(t), 0) = V(t)\mathbb{I}_{V(t)>0}. \tag{7.14}$$

By this decomposition, we can observe that in order to estimate the CVA payoff, we need to estimate the value of the underlying instrument $V(t)$ (that is typically much easier than the estimation of the value of the same instrument with optionality) and the indicator function $\mathbb{I}_{V(t)>0}$. In particular, for what concerns the latter expression, we observe that we need to be very accurate only when $V(t) \sim 0$; in all other cases, it is not really important to know the exact value of $V(t)$, as the indicator function maps it in the only two outcomes: zero or one. As a consequence, in order to speed up the calculation, one just needs to find an approximated version of $\mathbb{I}_{V(t)>0}$, $\mathbb{I}_{\hat{V}(t)>0}$, for example, using just a few runs of MC simulations. Given this approximation, one can rewrite Eq. (7.9) as

$$\text{CVA}_0 \sim (1 - R)\mathbb{E}^{\mathbb{Q}} \left[\int_0^T D(0,t)V(t)\mathbb{I}_{\hat{V}(t)>0}\delta(\tau - t)dt \right]. \tag{7.15}$$

Now, we have to remember the risk-neutral pricing formula,

$$V(t) = \mathbb{E}^{\mathbb{Q}}\left[D(t,T)\Phi(T)|\mathcal{F}(t)\right], \tag{7.16}$$

where $\Phi(T)$ is the payoff of the financial instrument. Substituting this expression in Eq. (7.15), we obtain

$$\text{CVA}(t = 0) \sim (1 - R)\mathbb{E}^{\mathbb{Q}} \left[\int_0^T D(0,t)\mathbb{E}^{\mathbb{Q}}\left(D(t,T)\Phi(T)|\mathcal{F}(t)\right) \right.$$

$$\left. \mathbb{I}_{\hat{V}(t)>0}\delta(\tau - t)dt \right]$$

$$= (1 - R)\mathbb{E}^{\mathbb{Q}} \left[\mathbb{E}^{\mathbb{Q}} \left(\int_0^T D(0,t)D(t,T)\Phi(T)|\mathcal{F}(t) \right) \right.$$

$$\left. \mathbb{I}_{\hat{V}(t)>0}\delta(\tau - t)dt \right]$$

$$= (1 - R)\mathbb{E}^{\mathbb{Q}} \left[\int_0^T D(0,T)\Phi(T)\mathbb{I}_{\hat{V}(t)>0}\delta(\tau - t)dt \right], \tag{7.17}$$

where in the last equation, we made use of iterated conditioning (Eq. (2.23)) and the definition of $D(x, y)$ (Eq. (3.8)) in order to simplify the product of the discount factors.

Equation (7.17) describes the same CVA problem from a different perspective; in this case, all we need to estimate is a *future* payoff multiplied by an approximation of the indicator function.

If from a theoretical point of view, the two approaches are equivalent, as a discounted future payoff multiplied by the indicator function of its positive value actually *is* the optional part of the CVA estimation, from a technological perspective, the implementation strategy of Eq. (7.17) can be remarkably better and it can provide different results in terms of computation time with respect to Eq. (7.13). In addition, one should keep in mind that the efficiency of the computational method, we choose to estimate the CVA, depends also on the complexity of the portfolio we are going to price. As a consequence, a unique representation could be not efficient in all the situations.

To conclude this section, we remark that the conditioning to the default time (Eq. (7.10)) could strongly affect the result of the CVA estimation. As a consequence, for some product, the assumption that the default time is independent of the optional part of the CVA could be quite strong, and an additional correction should be included into the CVA estimation. For further references on this topic, the interested reader can refer to Refs. [51–54]. In general, we remark that typically the default event is related to an abrupt change in the market conditions that can be hardly associated to a continuous hedging strategy.

7.4. Counterparty Credit Risk

In Chapter 6, we focused on the market risk estimation, i.e. the estimation of the risk of losing money because of the fluctuation of the market value of a given portfolio. In this chapter, we argue that this is not the only source of risk that could affect our portfolio. For example, one could be interested in the risk that a borrower will default on any type of debt by failing to make required payments; this kind of risk is called *Credit Risk*.

In this section, we deal with a more specific feature of Credit Risk, *Counterparty Credit Risk* (CCR) [55] and it is defined as [56] *the risk taken on by an entity entering an OTC contract with one (or more) counterparty having a relevant default probability. As such, the counterparty might not respect its payment obligation.*

In the previous section, we have already discussed about how to include the *premium* for this risk in the pricing algorithm, obtaining the so-called

CVA. In this section, we want to estimate the *risk* related to counterpart default.

Essentially, CCR differs from CVA for three main reasons:

- CCR, as a kind of risk, typically refers to extreme scenarios that could potentially affect our portfolio. On the contrary, CVA refers to *expected* future exposures (weighted by default probability). In mathematical terms, CCR could require the estimation of extreme percentiles, while CVA is based on the expected value.
- CCR is based on the physical measure as it does not refer to any hedging strategy. On the contrary, CVA refers to the pricing of an instrument, and it is based on the risk-neutral measure.
- CVA is used in practice as an adjustment to the price, CCR model is used to estimate the required capital to face losses because of counterparty defaults and to assign credit limit to each counterparty.

As a consequence, CCR methodologies can be quite different from CVA estimation. In particular, for what concerns the scenarios generation, CCR is usually based on a historical calibration while CVA requires a risk-neutral one.

Quite a classical approach for CCR estimation is to assume the three elements of Eq. (7.13), i.e. Loss-Given-Default (LGD), exposure at default (EAD), and the probability of default (PD) factor. So, the risk is simply given by the product of these three elements:

$$\text{CCR} = (L_{\text{GD}})(\text{EAD})(\text{PD}), \tag{7.18}$$

where each element is estimated in the physical measure. Also in this case, as in Eq. (7.13), we are neglecting the correlation between the three sources of risk, and each element can be estimated independently. In particular, usually banks and financial institutions have already available PD and L_{GD} relying on other models related to credit risk estimation, while the EAD is the real unknown of the CCR problem. We firstly remark that, as we are dealing with the physical measure, we are not required to define the exposure at default by the expectation operator as done in the CVA case; on the contrary, we can rely on a more general framework considering different definitions.

In order to deal with EAD estimation, as a first step, we observe that also in this case, we are interested in future exposures, so a time component must be included in the definition. As a consequence, one could fix some future time and consider the possible exposures as the functions of different market scenarios. For example, considering an equity forward (Section 3.3.4.3), one could fix a time horizon of one year and generate future values of the underlying assuming a lognormal distribution. Then, one could *reprice* the forward (using Eq. (3.24)) for each generated value of the underlying, obtaining the distribution of the future forward prices. Also in the case of CCR estimation, as in the CVA estimation, we are only interested in the *positive part* of the distribution that implies the application of the max operator. At this point, we just need to choose a suitable confidence level of the distribution that we think is representative of risk we want to describe, following the same philosophy we considered in the VaR estimation. In particular, as we are in the physical measure framework, we can choose an arbitrary confidence level to describe our EAD. In formulas, we can define the *Positive Future Exposure* (PFE) as

$$\alpha \equiv \int_0^{\text{PFE}(t_0,t_1)} \mathrm{d}F\left(\max(Y(t_1,T),0)\right), \tag{7.19}$$

where $Y(t_1,T)$ is the price at time t_1 of an equity forward with maturity T, α is the confidence level of our risk measure, and $\text{PFE}(t_0,t_1)$ is the Potential Future Exposure at time t_1 as seen at time t_0. We observe that in this situation, contrary to VaR case, we are dealing with a distribution defined only on the positive part of the x-axis.

In addition, in this case, the *larger* the exposure, the *bigger* the risk, so we have to focus on the right tail of the distribution, choosing the suitable confidence level, e.g. 99%.

On the same line, one can define a metric to obtain the *typical* value of the positive exposure,

$$\text{EE}(t_0,t_1) \equiv \mathrm{E}^{\mathbb{P}}\left(\max(Y(t_1,T),0)\right), \tag{7.20}$$

where $\text{EE}(x,y)$ stands for *expected exposure* at the time t_1 as seen at time t_0. Equation (7.20) is the physical measure $\mathbb{P}$ counterpart of the EAD we considered in risk-neutral framework (Section 7.3.1) to estimate CVA.

As evident, Eqs. (7.19) and (7.20) depend on the future time t_1 we choose to estimate the exposure. On the other side, we are interested in defining comprehensive risk measure that aims to estimate the counterparty risk, irrespective of the specific future time we choose for the measure. As a consequence, we can define the expected positive exposure (EPE), taking the time average of the EE,

$$\text{EPE}(t_0) \equiv \frac{1}{T} \sum_{i=1}^{n} \text{EE}(t_0, t_i), \qquad (7.21)$$

where T is the maturity of the instrument, and we considered a suitable partition of the time interval $[t_0, T]$ in n-parts.

As EE is related to expected scenarios, one could consider a slightly different definition to include the maximum of the past expected exposure in order to be more conservative. As a consequence, one can define the *effective expected exposure* (EEE) as

$$\text{EEE}(t_0, t_1) \equiv \max_{t < t_1} \left(\text{EE}(t_0, t_1) \right). \qquad (7.22)$$

We observe that, by definition, $\text{EEE}(t_0, t_1)$ is a non-decreasing function of t_1. Analogous to EPE, we can take the time-average of EEE in order to obtain the *effective expected positive exposure* EEPE,

$$\text{EEPE}(t_0) \equiv \frac{1}{T} \sum_{i=1} \text{EEE}(t_0, t_i). \qquad (7.23)$$

All the metrics introduced, especially EEPE and PFE, are of fundamental importance in the CCR framework, and they are also explicitly mentioned in regulatory documents as good proxies for risk estimation. In any case, these measures are not unique, and other approaches can be developed in order to estimate risk. For example, one could consider an expected shortfall approach as described in Section 6.4.

In any case, irrespective of the chosen measure, the main CCR task consists in the estimation of future positive exposures as a function of different risk-factor scenarios. From this point of view, CCR estimation is very similar to VaR estimation with the remarkable exception that VaR measure refers to a given time horizon (typically 1 day for managerial purposes and 10 days for regulatory purposes). On the contrary, in the CCR case, we need an exposure profile as a function of time, ranging over very long time horizons that depend on the maturities of the instruments in the portfolio. So, from this perspective, CCR estimation increases the complexity of VaR estimation, adding the *time dimension* to the problem. This has huge impact on the computational effort required. For example, if VaR calculation requires the revaluation of the whole portfolio over $n = 500$ scenarios, for the CCR estimation, assuming that the maximum maturity of the portfolio instruments is 10 years, the same calculation should be performed 10×365 times, assuming a daily revaluation of the exposure. In practice, these kinds of calculations are typically performed using suitable technical infrastructure, relying on parallel computing methods and optimized numerical simulations which allow risk estimation with reasonable computational times.

Given the structure of the problem, the typical approach to simulate risk factor dynamics is to rely on Monte Carlo simulations, starting from a (postulated) stochastic process with empirically calibrated parameters. Actually, this approach is quite compulsory, as empirical distributions over long time horizons are very difficult to be obtained,[c] so assumptions on the time scaling of the risk-factor distributions are required in order to generate future scenarios.

In order to minimize the counterparty risk, the two entities involved in the contract usually exchange money on a daily basis as a collateral of the exposure. In particular, every day (or at another agreed frequency) the entity with a positive exposure (the creditor) requires from its counterpart (the borrower) the payment of an equivalent amount of money (the collateral) in order to hedge counterparty risk. On the other hand, the creditor pays to the borrower an agreed rate (typically a proxy of the risk-free rate) on the posted money. As time passes, the exposure between the two counterparts can change because of price variation of the contract, as a consequence, the value of the collateral is adjusted accordingly.

Usually, the collateral interest rate, the payment frequency, and, in general, all the details about the collateral management are described in a legal document, the *Credit Support Annex* or *CSA* that regulates the credit support for derivative transactions.

The presence of collateral associated to financial contracts remarkably decreases the CCR as in the case of positive exposure toward a defaulted counterpart, the creditor can access collateral as a repayment of his credit. In this case, CCR is represented only by the part of the exposure that is not covered by collateral[d] that is typically a small percentage of the

---

[c]For example, a distribution of 500 observations over a time horizon of 1 year would require a 500-year historical series that is typically not available.

[d]As the collateral payment frequency is not continuous, some mismatch can be observed between the price of the instrument at default and the last paid collateral. These mismatches can be roughly estimated assuming normality and considering the price volatility of the instrument multiplied by the squared root of time delay between two payments.

total amount. From the modeling perspective, the collateral dynamics must be included in the CCR model in order to take into account correctly its effects on the risk measure and to assure its optimized management. This peculiarity can remarkably increase the complexity of the CCR model, requiring additional care with respect to the hypothesis applied and the computational performances.

Appendix

A.1. Black–Scholes Pricing Formula Derivation — Risk-Neutral Approach

We want to solve the following equation:

$$C(S,t) = \mathbb{E}^{\mathbb{Q}}\left[e^{-(T-t)r} \max\left(S(T) - K, 0\right) | \mathcal{F}(t) \right], \qquad \text{(A.1)}$$

assuming that $S(T)$ is lognormally distributed with parameters r and σ, as implied by risk-neutrality assumption. In the following, we will assume that $t = 0$. In other words, the $S(T)$ random variable is distributed according to the following probability density function (PDF):

$$p_{LN}(S(T)) = \frac{1}{\sqrt{2\pi\sigma^2 T}S(T)} e^{-\frac{\left[\log\left(\frac{S(T)}{S_0}\right) - \left(r - \frac{\sigma^2}{2}\right)T\right]^2}{2\sigma^2 T}}. \qquad \text{(A.2)}$$

Applying the integral representation of the expected value, we obtain that the integral to be solved is

$$C(S, t = 0) = \int_0^{+\infty} \max\left(S(T) - K, 0\right) p_{LN}(S(T)) \mathrm{d}S(T) \qquad \text{(A.3)}$$

$$= \int_K^{+\infty} \left(S(T) - K\right) p_{LN}(S(T)) \mathrm{d}S(T) \qquad \text{(A.4)}$$

$$= \int_K^{+\infty} S(T) p_{LN}(S(T)) \mathrm{d}S(T) \qquad \text{(A.5)}$$

$$- K \int_K^{+\infty} p_{LN}(S(T)) \mathrm{d}S(T), \qquad \text{(A.6)}$$

where in Eq. (A.5), we exploited the fact that for $S(T) < K$, the payoff is zero because of the *max* operator, and in Eq. (A.6), we split the integral in two parts that we are going to solve separately. In the following, we will drop the dependency by T of the variable $S(T)$ in order to simplify the notation.

We start with the second integral,

$$I_2 \equiv \int_K^{+\infty} \frac{1}{\sqrt{2\pi\sigma^2 T}S} e^{-\frac{\left[\log\left(\frac{S}{S_0}\right)-\left(r-\frac{\sigma^2}{2}\right)T\right]^2}{2\sigma^2 T}} dS, \qquad (A.7)$$

and apply the following change of variable:

$$z \equiv \frac{\log\left(\frac{S}{S_0}\right) - \left(r - \frac{\sigma^2}{2}\right)T}{2\sigma^2 T}. \qquad (A.8)$$

As a consequence, the Jacobian is

$$dz = \frac{1}{\sigma\sqrt{T}S} dS, \qquad (A.9)$$

and the integral becomes

$$I_2 = \int_{\frac{\log\left(\frac{K}{S_0}\right)-\left(r-\frac{\sigma^2}{2}\right)T}{2\sigma^2 T}}^{+\infty} \frac{1}{\sqrt{2\pi}} e^{-\frac{z^2}{2}} dz. \qquad (A.10)$$

We now define the normal CDF as

$$N(x) = \int_{-\infty}^{x} \frac{1}{\sqrt{2\pi}} e^{-\frac{u^2}{2}} du \qquad (A.11)$$

that is symmetric with respect to y-axis. So we obtain that

$$I_2 = N(d_2), \qquad (A.12)$$

where

$$d_2 = \frac{\log\left(\frac{S_0}{K}\right) + \left(r - \frac{\sigma^2}{2}\right)T}{2\sigma^2 T}. \qquad (A.13)$$

Now, we need to solve the first integral,

$$I_1 = \int_K^{+\infty} S p_{LN}(S) dS \qquad (A.14)$$

that by the change of variable $y = \log(S/S_0)$ becomes

$$I_1 = \int_{\log(K/S_0)}^{+\infty} \frac{S_0}{\sqrt{2\pi\sigma^2 T}} e^{-\frac{\left[y - \left(r - \frac{\sigma^2}{2}\right)T\right]^2}{2\sigma^2 T} + y} dy. \qquad (A.15)$$

Now, we define a new variable $\mu \equiv (r - \sigma^2/2)T$ and simplify the exponential part,

$$-\frac{\left[y - \left(r - \frac{\sigma^2}{2}\right)T\right]^2}{2\sigma^2 T} + y$$

$$= -\left[\frac{y^2 + \mu^2 - 2y\mu - 2y\sigma^2 T}{2\sigma^2 T}\right]$$

$$= -\frac{\mu^2}{2\sigma^2 T} - \left[\frac{y^2 + \mu^2 - 2(\mu + \sigma^2 T)y + (\mu + \sigma^2 T)^2 - (\mu + \sigma^2 T)^2}{2\sigma^2 T}\right]$$

$$= -\frac{\mu^2}{2\sigma^2 T} + \frac{(\mu + \sigma^2 T)^2}{2\sigma^2 T} - \frac{(y - (\mu + \sigma^2 T))^2}{2\sigma^2 T}$$

$$= \frac{\sigma^2 T}{2} + \mu - \frac{(y - (\mu + \sigma^2 T))^2}{2\sigma^2 T}. \qquad (A.16)$$

Substituting the result of our simplification in Eq. (A.15) and recollecting our definition of μ, we get

$$I_1 = \frac{S_0 e^{rT}}{\sqrt{2\pi\sigma^2 T}} \int_{\log(K/S_0)}^{+\infty} e^{-\frac{\left[y - \left(r + \frac{\sigma^2}{2}\right)T\right]^2}{2\sigma^2 T}} dy \qquad (A.17)$$

that, analogous to Eq. (A.10), gives the following result:

$$I_1 = S_0 e^{rT} N(d_1), \qquad (A.18)$$

where

$$d_1 \equiv \frac{\log\left(\frac{S_0}{K}\right) + \left(r + \frac{\sigma^2}{2}\right)T}{2\sigma^2 T} = d_2 + \sigma\sqrt{T}. \qquad (A.19)$$

Recovering the original expression of our problem, we get

$$C(S, t = 0) = SN(d_1) - Ke^{rT}N(d_2) \qquad (A.20)$$

that is the BS formula for the pricing of the call options.

A.2. Delta Formula for Call Options

In Section 5.2, we considered the following relation (Eq. (5.17)):

$$\Delta(t) = \frac{\partial C(S, t)}{\partial S(t)} = N(d_1) \tag{A.21}$$

that represents the amount of underlying needed to hedge a call option in the Black and Scholes framework. In this appendix, we obtain this equation, considering the derivative of the BS solution:

$$C(S, t = 0) = SN(d_1) - K e^{rT} N(d_2). \tag{A.22}$$

In this case, the differentiation is quite straightforward:

$$\frac{\partial C}{\partial S} = N(d_1) + S \frac{\partial N(d_1)}{\partial d_1} \frac{\partial d_1}{\partial S} - K e^{-rT} \frac{\partial N(d_2)}{\partial d_2} \frac{\partial d_2}{\partial S}. \tag{A.23}$$

In order to prove that Eq. (5.17) holds, we need to show that

$$S \frac{\partial N(d_1)}{\partial d_1} \frac{\partial d_1}{\partial S} - K e^{-rT} \frac{\partial N(d_2)}{\partial d_2} \frac{\partial d_2}{\partial S} = 0. \tag{A.24}$$

From the definitions of d_1 and d_2 in Eq. (5.10), we observe that

$$\frac{\partial d_1}{\partial S} = \frac{\partial d_2}{\partial S}, \tag{A.25}$$

so we can rearrange Eq. (A.24)

$$\frac{\partial d_1}{\partial S} \left(S \frac{\partial N(d_1)}{\partial d_1} - K e^{-rT} \frac{\partial N(d_2)}{\partial d_2} \right) = 0, \tag{A.26}$$

and we can focus on the term in the brackets. In particular, recollecting that

$$\frac{dN(x)}{dx} = \frac{e^{-\frac{x^2}{2}}}{\sqrt{2\pi}}, \tag{A.27}$$

we obtain

$$S \frac{e^{-\frac{d_1^2}{2}}}{\sqrt{2\pi}} - K \frac{e^{-\frac{d_2^2}{2}}}{\sqrt{2\pi}} = 0$$

$$e^{\frac{-d_1^2 + d_2^2}{2}} = \frac{K}{S} e^{-rT}. \tag{A.28}$$

Substituting in the exponential term the definitions of d_1 and d_2, one obtains

$$\frac{-d_1^2 + d_2^2}{2} = \log(K/S) - rT, \tag{A.29}$$

that implies that Eq. (A.28) is the identity.

A.3.   Heston Model Pricing Formula Derivation

We want to solve Eq. (5.77)

$$\left(-rC + \frac{\partial C}{\partial t} + rS\frac{\partial C}{\partial S} + k(\theta - \nu)\frac{\partial C}{\partial \nu} \right.$$
$$\left. + \frac{1}{2}S^2\nu\frac{\partial^2 C}{\partial S^2} + \frac{1}{2}\eta^2\nu\frac{\partial^2 C}{\partial \nu^2} + \eta\rho S\nu\frac{\partial^2 C}{\partial S\partial \nu} \right) dt = 0, \tag{A.30}$$

assuming that the solution is of the form (Eq. 5.79),

$$C(S, \nu, t) = S(t)P_1 - Ke^{-r(T-t)}P_2, \tag{A.31}$$

where P_j, $j = 1, 2$ are two unknown functions. We first define $x \equiv \log(S)$ and apply Itô's lemma (Eq. (2.93)),

$$\frac{\partial C}{\partial S} = \frac{\partial C}{\partial x}\frac{\partial x}{\partial S} = \frac{\partial C}{\partial x}\frac{1}{S},$$

$$\frac{\partial^2 C}{\partial \nu \partial S} = \frac{\partial}{\partial \nu}\left(\frac{\partial C}{\partial S} \right) = \frac{1}{S}\frac{\partial^2 C}{\partial \nu \partial x},$$

$$\frac{\partial^2 C}{\partial S^2} = \frac{\partial}{\partial S}\left(\frac{1}{S}\frac{\partial C}{\partial x} \right) = -\frac{1}{S^2}\frac{\partial C}{\partial x} + \frac{1}{S^2}\frac{\partial^2 C}{\partial x^2}. \tag{A.32}$$

Substituting them into Eqs. (A.30) and (A.31), we obtain

$$\frac{\partial C}{\partial t} + \frac{1}{2}\nu\frac{\partial^2 C}{\partial x^2} + \left(r - \frac{1}{2}\nu \right)\frac{\partial C}{\partial x} + \rho\eta\nu\frac{\partial^2 C}{\partial \nu \partial x}$$
$$+ \frac{1}{2}\eta^2\nu\frac{\partial^2 C}{\partial \nu^2} - rC + (k(\theta - \nu) - \lambda\nu)\frac{\partial C}{\partial \nu} = 0, \tag{A.33}$$

and

$$C(S, \nu, t) = e^x P_1 - Ke^{-r(T-t)}P_2. \tag{A.34}$$

We can now estimate all partial derivatives required by two-dimensional Itô's lemma (Eq. (2.101)),

$$\frac{\partial C}{\partial t} = e^x \frac{\partial P_1}{\partial t} - rKe^{-r(T-t)} P_2 - Ke^{-r(T-t)} \frac{\partial P_2}{\partial t},$$

$$\frac{\partial C}{\partial x} = e^x (P_1 + \frac{\partial P_1}{\partial x}) - Ke^{-r(T-t)} \frac{\partial P_2}{\partial x},$$

$$\frac{\partial^2 C}{\partial x^2} = e^x \left(P_1 + 2\frac{\partial P_1}{\partial x} + \frac{\partial^2 P_1}{\partial x^2} \right) - Ke^{-r(T-t)} \frac{\partial P_2}{\partial x^2},$$

$$\frac{\partial C}{\partial \nu} = e^x \frac{\partial P_1}{\partial \nu} - Ke^{-r(T-t)} \frac{\partial P_2}{\partial \nu},$$

$$\frac{\partial^2 C}{\partial \nu^2} = e^x \frac{\partial^2 P_1}{\partial \nu^2} - Ke^{-r(T-t)} \frac{\partial^2 P_2}{\partial \nu^2},$$

$$\frac{\partial^2 C}{\partial x \partial \nu} = e^x \left(\frac{\partial P_1}{\partial \nu} + \frac{\partial^2 P_1}{\partial \nu \partial x} \right) - Ke^{-r(T-t)} \frac{\partial^2 P_2}{\partial \nu \partial x}. \tag{A.35}$$

We can now substitute in Eq. (A.33), and we obtain

$$e^x \frac{\partial P_1}{\partial t} - rKe^{-r(T-t)} P_2 - Ke^{-r(T-t)} \frac{\partial P_2}{\partial t}$$

$$+ \frac{1}{2}\nu \left\{ e^x \left(P_1 + 2\frac{\partial P_1}{\partial x} + \frac{\partial^2 P_1}{\partial x^2} \right) - Ke^{-r(T-t)} \frac{\partial P_2}{\partial x^2} \right\}$$

$$+ \left(r - \frac{1}{2}\nu \right) \left\{ e^x \left(P_1 + \frac{\partial P_1}{\partial x} \right) - Ke^{-r(T-t)} \frac{\partial P_2}{\partial x} \right\}$$

$$+ \rho\eta\nu \left\{ e^x \left(\frac{\partial P_1}{\partial \nu} + \frac{\partial^2 P_1}{\partial \nu \partial x} \right) - Ke^{-r(T-t)} \frac{\partial^2 P_2}{\partial \nu \partial x} \right\}$$

$$+ \frac{1}{2}\eta^2\nu \left\{ e^x \frac{\partial^2 P_1}{\partial \nu^2} - Ke^{-r(T-t)} \frac{\partial^2 P_2}{\partial \nu^2} \right\}$$

$$- r \left\{ e^x P_1 - Ke^{-r(T-t)} P_2 \right\}$$

$$+ (k(\theta - \nu) - \lambda\nu) \left\{ e^x \frac{\partial P_1}{\partial \nu} - Ke^{-r(T-t)} \frac{\partial P_2}{\partial \nu} \right\} = 0. \tag{A.36}$$

Now, we observe that Eq. (A.33) must hold for all combinations of S, K, r. This fact allows us to split the problem into two equations, one for

P_1 imposing $K = 0, S = 1$ and one for P_2, imposing $S = 0, K = 1, r = 0$ (see Eq. (A.31)). Using this method, the problem described by Eq. (A.36) can be reduced by two equations for P_1, P_2,

$$\frac{\partial P_j}{\partial t} + \rho\eta\nu\frac{\partial^2 P_j}{\partial x \partial \nu} + \frac{1}{2}\nu\frac{\partial^2 P_j}{\partial x^2} + \frac{1}{2}\eta^2\nu\frac{\partial^2 P_j}{\partial \nu^2}$$

$$+(r + u_j\nu)\frac{\partial P_j}{\partial x} - (a - b_j\nu)\frac{\partial P_j}{\partial \nu} = 0 \qquad (A.37)$$

for $j = 1, 2$ where $u_1 = \frac{1}{2}$, $u_2 = -\frac{1}{2}$, $a = k\theta$, $b_1 = k - \rho\eta$ and $b_2 = k$. Equation (A.37) must satisfy the terminal condition implied by Eq. (5.78),

$$P_j(x, \nu, T; \log(K)) = \mathbb{I}_{x \geq \log(K)}. \qquad (A.38)$$

Now, we want to show that the function P_j can be interpreted as the probability for a suitable stochastic process $X(t)$ of being larger than (the logarithm of) K at time T as required by the condition of Eq. (A.38). We first consider two Stochastic Differential Equations (SDEs),

$$dX_j = (r + u_j\nu)dt + \sqrt{\nu}dW_1,$$

$$d\nu_j = (a - b_j\nu)dt + \eta\sqrt{\nu}dW_2. \qquad (A.39)$$

and the function,

$$h(X(t), \nu(t), t) = \mathrm{E}\left[g(X(T), \nu(T))|\mathcal{F}(X(t), \nu(t), t)\right], \qquad (A.40)$$

where $g(x, y)$ is a generic function, and we denoted the $\mathcal{F}(X(t), \nu(t), t)$ the filtration generated by $X(t)$ and $\nu(t)$ at time t. Applying the iterated conditioning property of the expectation operator, it can be proved that $h(X(t), \nu(t), t)$ is a martingale. As a consequence, the drift term of its related SDE must be equal to zero. Applying Itô's lemma, we obtain

$$dh = \left[\frac{\partial h}{\partial t} + (r + u_j\nu)\frac{\partial h}{\partial x} + (a - b_j\nu)\frac{\partial h}{\partial \nu}\right.$$

$$\left. + \frac{1}{2}\nu\frac{\partial^2 h}{\partial x^2} + \frac{1}{2}\eta^2\nu\frac{\partial^2 h}{\partial \nu^2} + \rho\eta\nu\frac{\partial^2 h}{\partial x \partial \nu}\right]dt$$

$$+ [\ldots]dW_1 + [\ldots]dW_2, \qquad (A.41)$$

and equating the drift part to zero, we obtain again Eq. (A.37). In other words, as a consequence of the Feynman–Kac theorem, we could transform our original PDE problem into an expectation estimation problem, Eq. (A.40), where the SDEs of the process are given by Eq. (A.39), and the final condition can be expressed by a suitable choice of the function $g(x, y)$, i.e.

$$P_j = \mathbb{E}\left[\mathbb{I}_{X_j \geq \log(K)} | \mathcal{F}(X(t), \nu(t), t)\right] = \mathbb{Q}\left(X_j \geq \log(K)\right). \tag{A.42}$$

By the last expression, we want to describe the probability $\mathbb{Q}$ for X_j of being larger or equal than the logarithm of K. This proves our interpretation of P_j in probability terms. Unfortunately, it is not possible to obtain an analytical expression for the probabilities P_j, and we need to express them in terms of their characteristic function,

$$f_j(X, \nu, t, \psi) = \mathbb{E}\left[e^{i\psi X(t)}\right]. \tag{A.43}$$

By the new choice of the function $g(X, \nu) \equiv e^{iX\psi}$, we can observe that the equation, that our characteristic function must satisfy, is exactly equal to the one for P_j, i.e. Eq. (A.41), except for the final condition. Assuming that f_j has the form,

$$f_j(X, \nu, t; \psi) = e^{C_j(T-t,\phi)+D_j(T-t,\psi)+i\psi X}, \tag{A.44}$$

where C_j and D_j are two unknown functions, we can obtain their equations substituting Eq. (A.44) into Eq. (A.41). It turns out for the characteristic function the equations have the following form:

$$\frac{\partial D_j}{\partial \tau} = \rho \eta i \psi D_j - \frac{1}{2}\psi^2 + \frac{1}{2}\eta^2 (D_j)^2 + u_j i\psi - b_j D_j,$$

$$\frac{\partial C_j}{\partial \tau} = ir\psi + aD_j, \tag{A.45}$$

where $\tau \equiv T - t$. From the terminal condition $f_j(X, \nu, T; \psi) = e^{iX(T)\psi}$, we obtain the initial conditions $D_j(\tau = 0) = 0$ and $C_j(\tau = 0) = 0$. The first of Eq. (A.45) is a Riccati equation in D_j, while the second one can be solved using a straightforward integration once D_j is known. The solutions of these equations are given by Eqs. (5.82) and (5.83). Once f_j is obtained, we can invert them determining P_j as given by Eq. (5.80).

Bibliography

[1] S. E. Shreve, *Stochastic Calculus for Finance. Volume I — The Binomial Asset Pricing Model*. Volume II — *Continuous Time Models*, Springer, New York, 2004.

[2] R. Cont, P. Tankov, *Financial Modelling with Jump Processes*, CRC Press, 2003.

[3] W. Feller, *An Introduction to Probability Theory and its Applications*, Vols. 1 and 2, Wiley, New York, 1950–1966.

[4] E. J. Gumbel, *Statistics of Extremes*, Dover Publications, Mineola, NY, 2004.

[5] B. K. Oksendal, *Stochastic Differential Equations: An Introduction with Applications*, Springer, Berlin, 2003.

[6] K. Itô, *Stochastic Integral, Proceedings of the Imperial Academy*, 20(8), 519–524, 1944.

[7] M. Musiela, M. Rutkowski, *Martingale Methods in Financial Modelling*, Springer, New York, 2004.

[8] F. Black, M. Scholes, The pricing of options and corporate liabilities, *Journal of Political Economy*, 81, 637–654, 1973.

[9] E. Derman, The boy's guide to pricing and hedging, *Risk Magazine*, 16, 70–72, 2003.

[10] V. V. Piterbarg, Funding beyond discounting: Collateral agreements and derivatives pricing, *Risk*, 24, 97–102, 2010.

[11] V. V. Piterbarg, Cooking with collateral, *Risk Magazine*, 25(8), 58–63, 2012.

[12] D. Brigo, A. Pallavicini, *CCP Cleared or Bilateral CSA Trades with Initial/Variation Margins Under Credit, Funding and Wrong-way Risks: A Unified Valuation Approach*, preprint, 2014 arXiv:1401.3994.

[13] J. M. Harrison, S. R. Pliska, Martingales and stochastic integrals in the theory of continuous trading, *Stochastic Processes and their Applications*, 11, 215–260, 1981.

[14] J. M. Harrison, S. R. Pliska, A Stochastic calculus model of continuous trading: Complete markets, *Stochastic Processes and their Applications*, 15(3), 313–360, 1983.

[15] R. C. Merton, Theory of rational option pricing, *Bell Journal of Economics and Management Science*, 4, 141–183, 1973.

[16] H. Buehler, Statistical hedging - cost, carry, risk, *Global Derivatives Trading & Risk Management Conference*, Amsterdam, 2014.

[17] H. Buehler, Statistical hedging: Application to stochastic local volatility models, *Global Derivatives Trading & Risk Management Conference*, Amsterdam, 2013.

[18] H. Buehler, Statistical Hedging with Stochastic Local, *MathFinance Conference*, Frankfurt, 2013.

[19] D. T. Breeden, R. H. Litzenberger, Prices of state-contingent claims implicit in option prices, *The Journal of Business*, 51(4), 621–651, 1978.

[20] L. Spadafora, G. P. Berman, F. Borgonovi, Adiabaticity conditions for volatility smile in Black–Scholes pricing model, *European Physical Journal B*, 79, 47–53, 2010.

[21] S. L. Heston, A Closed-form solution for options with stochastic volatility with applications to bond and currency options, *The Review of Financial Studies*, 6(2), 327–343, 1993.

[22] J. C. Cox, J. E. Ingersoll, S. A. Ross, A Theory of the term structure of interest rates, *Econometrica*, 53(2), 385–407, 1985.

[23] Y. Maghsoodi, Solution of the extended CIR term structure and bond option valuation, *Mathemaical Finance*, 6(1), 89–109, 1996.

[24] D. Brigo, F. Mercurio, A deterministic-shift extension of analytically tractable and time-homogeneous short rate models, *Finance and Stochastics*, 5, 369–388, 2001.

[25] C. Kahl, P. Jackel, Not-so-complex logarithms in the Heston model, *Wilmott Magazine*, 19(9), 74–103, 2005.

[26] A. A. Dragulesco, V. M. Yakovenko, Probability distribution of returns in the Heston model with stochastic volatility, *Quantitative Finance*, 2(6), 443–453, 2002.

[27] C. Acerbi, Spectral measures of risk: A coherent representation of subjective risk aversion, *Journal of Banking and Finance*, 26(7), 1505–1518, 2002.

[28] P. Artzner, F. Delbaen, J. M. Eber, D. Heath, Coherent measures of risk, *Mathematical Finance*, 9(3), 203–228, 1999.

[29] R. Cont, R. Deguest, G. Scandolo, Robustness and sensitivity analysis of risk measurement procedures, *Quantitative Finance*, 10(6), 593–606, 2010.

[30] M. J. Schervish, *Theory of Statistics*, Springer Series in Statistics, Springer, 1996.

[31] J.-P. Bouchaud, M. Potter, *Theory of Financial Risk — From Statistical Physics to Risk Management*, Cambridge University Press, Cambridge, 2000.

[32] P. Embrechts, S. I. Resnick, G. Samorodnitsky, Extreme value theory as risk management tool, *North American Actuarial Journal*, 3(2), 30–41, 1999.

[33] A. F. McNeil, R. Frey, P. Embrechts, Quantitative Risk Management: Concepts, *Techniques and Tools*, Princeton University Press, princeton, 2015.

[34] C. Mancini, Non-parametric threshold estimation for models with stochastic diffusion coefficient and jumps, *Scandinavian Journal of Statistics*, 36(2), 270–296, 2009.

[35] C. Mancini, V. Mattiussi, R. Renò, *Spot Volatility Estimation Using Delta Sequences*, DiMaD Working Paper, No 2012-10, Università degli Studi di Firenze, 2012.

[36] J. C. Hull, A. White, Incorporating volatility updating into the historical simulation method for Value-at-Risk, *Journal of Risk*, 1(1), 5–19, 1998.

[37] P. Glasserman, *Monte Carlo Methods in Financial Engineering*, Springer Science and Business Media, Springer, 2004.

[38] P. Jäckel, *Monte Carlo Methods in Finance*, Wiley Finance, Chichester, 2002.

[39] J. Berkowitz, Testing the accuracy of density forecasts, applications to risk management, *Journal of Business & Economic Statistics*, 19(4), 465–474, 2001.

[40] F. Anfuso, D. Karyampas, A. Nawroth, Credit exposure models backtesting for Basel III, *Risk Magazine*, 26, 82–87, 2014.

[41] C. Acerbi, B. Szkely, Backtesting expected shortfall, *Risk Magazine*, 1–37, December, 2014.

[42] Basel Commitee on Banking Supervision, *Compilation of Documents that Form the Global Regulatory Framework for Capital and Liquidity*, Bank of International Settlements, available at http://www.bis.org/bcbs/basel3/compilation.htm.

[43] European Union, Regulation (EU) No 575/2013 on *Prudential Requirements for Credit Institutions and Investment Firms (CRR)*, http://ec.europa.eu/finance/bank/regcapital/legislation-in-force/index_en.htm, 2013.

[44] European Union, *Directive 2013/36/EU on Access to the Activity of Credit Institutions and the Prudential Supervision of Credit Institutions and Investment Firms (CRD IV)*, available at http://ec.europa.eu/finance/bank/regcapital/legislation-in-force/index_en.htm, 2013.

[45] Basel Commitee on Banking Supervision, *Minimum Capital Requirements for Market Risk*, Bank of International Settlements, 2016.

[46] M. Bianchetti, M. Carlicchi, *Interest Rates After the Credit Crunch: Multiple Curve Vanilla Derivatives and SABR*, available at SSRN: http://ssrn.com/abstract=1783070, 2011.

[47] M. Bianchetti, M. Morini (eds.), *Interest rate modelling after the financial crisis*, *Risk Books*, London, 2013.

[48] A. Pallavicini, D. Brigo, *Interest-rate Modelling in Collateralized Markets: Multiple Curves, Credit-liquidity Effects, CCPs*, available at SSRN: http://ssrn.com/abstract=2244580, 2013.

[49] International Accounting Standards Board, *IFRS 13, Fair Value Measurement*, 2011.

[50] J. Andreasen, *CVA on iPad Mini*, Global Derivatives, ICBI, Amsterdam, 2014.

[51] F. Mercurio, M. Li, *Jumping with Default: Wrong-way Risk Modelling for CVA*, available at SSRN: http://ssrn.com/abstract=2605648, 2015.

[52] D. Brigo, M. Morini, A. Pallavicini, *Counterparty Credit Risk, Collateral and Funding with Pricing Cases for All Asset Classes*, Wiley, 2013.

[53] D. Brigo, A. Pallavicini, *Counterparty Risk under Correlation between Default and Interest-Rates*, In Numerical Method for Finance, ed. Miller, Edelman, Appleby, CRC Financial Mathematics Series, Chapman & Hall, 2007.

[54] C. Burgard, M. Kjaer, PDE representations of options with bilateral Counterparty risk and funding costs, *Journal of Credit Risk*, 7(3), 1–19, 2011.

[55] J. Gregory, *Counterparty credit Risk and Credit Value Adjustment*, 2nd edn. Wiley Finance, Chichester, 2013.

[56] D. Brigo, *Counterparty Risk FAQ: Credit VaR, PFE, CVA, DVA, Closeout, Netting, Collateral, Re-hypothecation, WWR, Basel, Funding, CCDS and Margin Lending*, preprint, 2011, available at arXiv.org, http://arxiv.org/abs/1111.1331.

Index